MW00475553

Stories From My Sensei

Two Decades of Lessons
Learned Implementing
Toyota-Style Systems

Advance Praise for *Stories from My Sensei*

"Steve is a real sensei. His presentations captivate students. His enthusiasm is contagious. When Steve writes about the TPS house it is not just an abstraction. It is something he lives. He can vividly picture it in every operation. He can guide the organization through the journey to build a real system. You will enjoy Steve's stories that will bring the house to life."

—Jeffrey K. Liker, professor and author of *The Toyota Way*

"Steve Hoeft has written an insightful and highly entertaining memoir describing his lean journey. He recounts powerful stories drawn from his work career to illustrate how an experienced Toyota sensei mentored his development. Hoeft uses the framework of the Toyota Production System house to integrate the narrative, with examples that make Toyota philosophy, methods, and tools spring to life."

—John "Jack" Billi, M.D., Professor-Internal Medicine and Medical Education, Assoc. Dean-Clinical Affairs, Medical School, Assoc. VP-Medical Affairs, University of Michigan Health System

"In *Stories from My Sensei* you will not only learn the principles that underscore TPS, but will learn them through hands-on stories that anyone involved in operations and lean implementation can directly relate to. In short, it will make you laugh and cry; knowing that true learning is accomplished through the joy of suffering!"

—Jim Huntzinger, President and Founder of the Lean Accounting Summit and TWI Summit (www.leansummits.com) and author of *Lean Cost Management: Accounting for Lean by Establishing Flow*

"Powerful and entertaining examples that bring key Lean concepts to life. You will see Steve's lively and inspired style coming through the pages---just as one would expect from a very knowledgeable and masterful Lean practitioner, storyteller, and instructor. His participation in the University of Michigan Lean program as an instructor has set the bar high for all of us and generated high praise from many past Lean program participants as evidenced by some of the actual quotes below:

- The instructor had a lot of practical examples that would apply to our company.
- Outstanding experience and sharing.
- Outstanding! Steve is an outstanding instructor/facilitator, the animation and real-life stories make the course fun. Very excited and passionate.
- Great stories: please write your book Steve!
- Great job Steve!! The concepts will help tremendously.
- I can take this information to use in my Kaizen events.
- Steve is awesome!
- Excellent use of stories.
- Very entertaining and great examples. Very good knowledge. Thank you!
- Steve is excellent. Best instructor/teacher I have seen in years."

—Yavuz A. Bozer, Professor-Industrial & Operations Engineering, Co-Director, Tauber Institute for Global Operations, Co-Founder & Co-Director, Lean Manufacturing Certificate Program, University of Michigan

"Steve is one of the few leaders in the US that is truly a TPS expert. He has not only written the way to be successful but he also demonstrated it successfully within various companies while keeping TPS pure and simple."

—Phil Lardiere, President Pendleton Group Inc.

Stories From My Sensei

Two Decades of Lessons Learned Implementing Toyota-Style Systems

Steve Hoeft

Foreword by Jeffrey K. Liker

CRC Press
Taylor & Francis Group
Boca Raton London New York

CRC Press is an imprint of the
Taylor & Francis Group, an **informa** business

A PRODUCTIVITY PRESS BOOK

Productivity Press
Taylor & Francis Group
270 Madison Avenue
New York, NY 10016

© 2010 by Taylor and Francis Group, LLC
Productivity Press is an imprint of Taylor & Francis Group, an Informa business

No claim to original U.S. Government works

Printed in the United States of America on acid-free paper
10 9 8 7 6 5 4 3 2 1

International Standard Book Number: 978-1-4398-1654-7 (Hardback)

Library of Congress Cataloging-in-Publication Data

Hoeft, Steven E.
　　Stories from my sensei : two decades of lessons learned implementing Toyota-style systems / Steven E. Hoeft.
　　　　p. cm.
　　Includes bibliographical references and index.
　　ISBN 978-1-4398-1654-7 (hbk. : alk. paper)
　　1. Production management. 2. Toyota Jidosha Kabushiki Kaisha--Management. 3. Industrial efficiency. 4. Quality control--Management. 5. Industrial management. I. Title.

TS155.H576 2010
658.5--dc22   2009031548

Visit the Taylor & Francis Web site at
http://www.taylorandfrancis.com

and the Productivity Press Web site at
http://www.productivitypress.com

Dedication

To Gena, Megan, Erich and Erin. You mean more to me than the whole world.

Contents

Foreword

I had the honor of working with Steve Hoeft when he joined my company, Optiprise, as a senior lean consultant. When he showed me what he had done in leading the creation of a lean warehouse I was hooked, and we offered him the job immediately. His ability to not only *talk* Toyota Production System (TPS), but to actually *do* it, was what sold me.

Steve was taught by some of the original Japanese sensei when he worked for Johnson Controls, Inc. (JCI) in the seat division. JCI had won the contract to supply seats to Toyota's plant in Georgetown, Kentucky. As usual, Toyota was not satisfied simply to buy seats at a preset price but wanted to ensure a reliable supply of the highest quality seats. The only way they could ensure that was to teach JCI TPS. Some of the best Japanese experts taught JCI using the Toyota method of hands-on learning at the gemba. They were very successful making the plant a showcase for building seats just in time and delivering them in sequence to Toyota. Steve was tasked with taking what JCI learned in the Lexington, Kentucky plant, creating JCI's version of TPS, and teaching it to other parts of the company. Unfortunately, as is too often the case, the rest of the company was not all that committed to learning it and never came up to the same level.

Steve moved on and worked for several companies, including an engineering company that saw TPS as a tool to sell engineering services. Steve became the token lean guy. When he joined me I believe it was very liberating for him. He could actually teach TPS to some companies that cared—and others that did not so much.

Steve and I were in complete agreement that TPS is a system. That is why it is represented as a house. A house is a structural system and any weak part—the foundation, the structural supports, the roof—will make the whole house weak. Thinking about implementing a system sounds kind of cool, but it is actually a very scary proposition. You cannot shortchange a system by implementing just pieces of it—it is all or nothing. The challenge is that you also cannot implement a system all at once. It must be built in pieces but continued until you achieve at least a rudimentary level of each part. One of the key parts that is often neglected is "flexible, capable, highly motivated people." This takes far more time and patience than deploying tools like kanban.

You cannot just imitate Toyota piece by piece—a little kanban here, a cell there, put in the andon, develop standardized work, train the people, and let it rip. This will always fail. It is a living system and each part affects the other. Ohno learned that when the process is stable you can remove inventory, which makes the likelihood of shutting down production much greater, which requires more stability, which requires people to solve problems, which develops their capability, and on and on. This requires an environment of respect for people and trust.

Steve and I have worked with many companies who have started the journey to lean very enthusiastically and then at some point it fizzled. After getting some quick wins, senior management would lose interest and move on to the next fad. Before we knew it, the "system" was just individual pieces that were degrading over time.

Steve is a real sensei. His presentations in courses captivate students. His enthusiasm is contagious. He can run a kaizen event as well as the best, and the team will always have a remarkable report out, all beaming with pride. But any sensei needs a star pupil, which is the Holy Grail for a sensei. It is an endless search for the star student who listens, practices, learns, improves, and eventually surpasses the teacher. Unfortunately the organization that stays committed to lean and to learning long term is all too rare. Usually, changes in senior management or an economic downturn kills the momentum, and the house collapses before it is ever completely built.

When Steve writes about the TPS house, it is not just an abstraction. It is something he lives. He can vividly picture it in every operation. He can guide the organization through the journey to building a real system, as long as the senior leadership is committed. That is the part we cannot really control.

You will enjoy Steve's stories, which bring the house to life. These are all true stories and I know Steve treasures every one of them. It is what makes his lectures so captivating. I believe this book will captivate you as well. Enjoy the stories and learning what lean as a system really means, and perhaps you will be fortunate enough to work in an organization that makes a real long-term commitment to the vision.

Jeffrey K. Liker, PhD
Author, *The Toyota Way*

Prologue

Stories are a powerful vehicle for learning. Stories are memorable. And, stories tell a story. As a long-term instructor for the University of Michigan's Lean Certification programs, I had no idea how powerful and lasting stories could be. Even after several years, participants still go out of their way to repeat back to me parts of their favorite story that I told during a course. Instead of a cheery hello, I get, "Steve-san, stand in circle!" Oh well, they could have remembered one of the many big mistakes on my two-decade journey to learn all-things-Toyota.

This book weaves the story of one unassuming person's career through some powerful learning points. In this case, that career is mine. Some names have been changed to protect the innocent. And, some stories have been edited to bring out a more powerful lesson.

This book is organized around the Toyota House model for implementation. There is tremendous value in using a framework like the Toyota House to reinforce learning. I also maintain time order for the stories, with just a few exceptions, in each chapter of this book. Some have a name in parentheses before the story name. This tells you which employer I was working for at the time. The number in parentheses next to each story title represents the actual chronological order, for those who prefer to read them sequentially, and to better frame the story. The following list shows the chronological order of my employers:

- Admiral Engines (AE, automobile assembler)
- Glass Company
- Triangle Kogyo (complete seat and parts maker for Mazda)
- Johnson Controls, Inc. (JCI, complete seat and parts maker for all automobile assemblers)
- Consultant (applying Toyota principles, much of the time with Jeff Liker's firm Optiprise)
- Altarum Institute (nonprofit research and innovation institute)

Each chapter opens with a thought-provoking quote and a graphic of the Toyota House model showing the reader where they are (like the "you are here" star on a map at a shopping mall). These icons will help keep you focused on the sequence and importance of the model. The continuum is as follows:

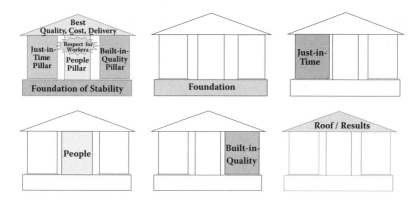

Each major part of the house is described in some detail, including some interesting historical notes. It is important to note the origins and original application in describing the tools and principles. Following each introduction, there will be a dozen or so stories from my sensei and other leaders that probably teach more about how to (and sometimes how not to) implement the Toyota Production System (TPS) than any lengthy section of writing ever could.

As we dig deeply into the foundation or base tools, you may be able to see me *learn*. You can learn from my many mistakes. You may have made only a dozen or so key mistakes in your career thus far. I may have made my 1000th dumb mistake just last week. I have learned to learn from each one. How about you?

Toyota leaders taught me the powerful technique of hansei, or reflection. I have participated in several hansei sessions. They used the past to redirect the future in a safe and non-accusatory way. After each story, I have listed a few questions. I encourage you to answer these questions with a critical eye and a willingness to recognize the gap between where you are and where you need to be. There is some sage advice in your own answers. Learn from your own learnings. And, share them so that others do not fall into the same traps. Write your thoughts down immediately as you think of them. Don't wait. Don't edit them. Read on. Then, read through these questions and your answers again after you have finished this book. Are you ready to learn?

Acknowledgments

This book summarizes over 20 years of pursuit of all-things-Toyota. Much of what I know, and teach, today was gleaned from the mouths and through the eyes of a dozen or more senseis, or master trainers of the Toyota Production System (TPS) principles. There is no way to thank every person who added knowledge to this toolkit and framework that even today is still growing. Unfortunately, I cannot acknowledge all of the individuals at Toyota Motor Manufacturing Kentucky, New United Motor Manufacturing Inc. (NUMMI), Toyota Technical Center, Johnson Controls, Delta Kogyo and several sensei-trained individuals that fueled the fires of TPS in me. But, several people were noteworthy on my journey and in writing this book.

My two earliest senseis brought me from grasshopper status to being almost dangerous in my ability to properly deploy TPS principles in the few short years that I was under their guidance. They also added a humorous element to the stories that you will read (although it was not humorous at the time). Since it has been over 15 years since my last contact with them, I have chosen to use pseudonyms rather than their real names throughout the book.

Dr. Jeffrey K. Liker—professor of Industrial and Operations Engineering at the University of Michigan, co-founder of multiple Lean Certification programs at the university, and one prolific writer—gave me a start in the stand-up arena teaching Toyota principles. From my first-ever delivery to a class of senior engineers at the university to regular teaching in Lean Certification programs, Jeff saw value in my stories and encouraged me to weave them into the lectures. Later, I would join Jeff's consulting company, Optiprise, where we had a great deal of fun implementing TPS principles at companies that told us, "It won't work here." If we only had a nickel for every time we heard that comment! Thanks, Jeff.

Two others I met through the University of Michigan were also crucial in correcting my Big Three–influenced thinking—John Shook and Mike Rother. John Shook first worked with Toyota in Japan, then became part of their story here in the United States. He and Mike Rother wrote *Learning to See** (as well as other books). John's insistence about the consistency and "purity" of TPS was well received. I was able to work with John and his staff on a couple of key projects applying TPS in manufacturing and healthcare. Mike Rother introduced me to principals at the Altarum Institute, where I work now. His passion for making things flow and visualizing value streams added much to my knowledge base.

Some other peers added to my TPS knowledge and stories as well. I was blessed to work with two committed TPS coaches, Keith Leitner and Dr. John Drogosz, while at Optiprise. I was also able to work with and learn from Dr. Jim Morgan and Jeff Liker in developing a Lean Product Process Development Certification program for the Society of Automotive Engineers, then later for the University of Michigan. Jim's three-year comparison of Toyota versus a North American auto assembler's new product development system led to a coherent systems model of lean product development. Jim co-authored *Toyota Product Development*† with Jeff Liker.

Two individuals from Altarum Institute helped with reviews and edits. Brock Husby (PhD, we hope, by the time this book is published) helped review this book and has added a lot of building blocks to the

---

* Rother, Michael and John Shook, Learning to See, Lean Enterprise Institute, 1999.
† Morgan, James M. and Jeffrey K. Liker, The Toyota Product Development System, Productivity Press, 2006.

growing knowledge base for TPS principles applied to healthcare. My boss, Cathy Call, also provided significant edits and reviews.

Maura May, publisher at Productivity Press/Taylor & Francis Group, was instrumental in leading me through a major rewrite of this book. She encouraged me throughout the process and offered great ideas.

One final acknowledgment is due. To my wife, Eugena Dee (Cowden) Hoeft—thank you for being a great wife, our kids' mom, and for putting up with a man engrossed in his work, church and life. What would I do without you?

SUMMARY OF THE TOYOTA HOUSE MODEL FOR IMPLEMENTATION

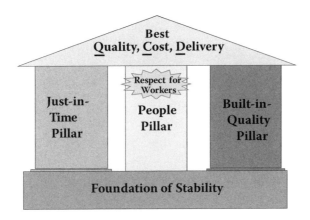

Has anyone ever asked you to describe your company's production system—or simply—the way you make stuff every day? If so, how would you answer? Would you start by describing the products, or maybe an organization chart? Would you dive into the details of the processes that are used?

If you ask a Toyota associate to describe the Toyota Production System (TPS), they will not start with any of the suggestions listed above. More than likely, they will whip out a blank piece of paper and draw for you what is commonly known as the "Toyota House" model. If they do not draw it, they will describe the principles in words that have some sort of sequence or framework. This actually happened during my first-ever meeting with a Toyota leader. It led to one of the most fantastic journeys I have ever undertaken.

This strange behavior of drawing a simple house structure with pillars and a roof is so consistent that it seems to flow out of Toyota associates' DNA. Maybe it *is* the DNA of Toyota, the long-sought-after Holy Grail. But, where did the house come from? And, why is it so important?

As Toyota began to spread its fledgling production system outside of Toyota City and Japan, they needed a way to quickly disseminate, communicate, and train suppliers and workers on the important parts of their best practices. It was a young Fujio Cho who first drew a house model to communicate these truths. (Cho-san later led Toyota's U.S. operations and then became company president.)

The Toyota House model still stands as the best simple representation of the principles and philosophy of TPS. A house is a great analogy for the truths that Toyota learned through constant trial and error over decades of producing textiles and then motor vehicles. A house is a good model both to show the sequence and building blocks *during construction*, as well as the durability and physical parts of its structure *after it has been built*. A stable house is strong and lasting.

TPS builds itself on a foundation of operational stability as shown in Figure 1.1 below. One of the main shortcomings of change programs of the past was that the tools or concepts somehow became the goal. In TPS transformations, implementing the tools is NOT the goal. Toyota leaders never forgot that they were building *something*. That something is a strong enterprise—not tools. It is represented well by a house.

A house has major parts like a foundation, walls or pillars, and a roof. In our Toyota House analogy, we put additional words inside

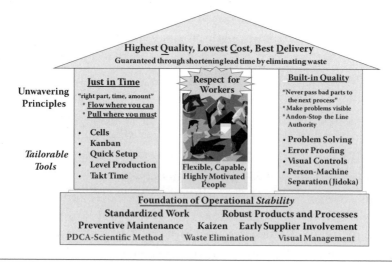

Figure 1.1   Toyota House Model, adapted from Toyota Motor Manufacturing, Kentucky.

these major parts. The major parts are required. They are structural and integral to making a house stand. The words inside the major parts are also required. We call these *principles*. The principles are timeless, unchanging and always implemented with thoughtfulness and determination.

There are also some words inside the pillars that represent "tools." There are 20 or so consistently used tools in the TPS (like workplace organization using 5S, kanban or one-piece flow cells). These tools are often applied in unique and tailored ways within an organization. The tools listed in the house were selected due to their importance and placed where they are typically first used. These will be noted with bullet symbols.

Many variations of this house model exist. The sequence of the pillars changes and more or fewer words appear in the boxes. Followers of Taiichi Ohno would probably label the right pillar "Jidoka." But, the timeless, unchanging, unwavering principles remain. The Toyota House model stands as a great representation and an easy-to-follow, effective framework for understanding these principles.

Building a house is a common analogy that many can follow. It demonstrates not only the importance of a good, solid foundation, but also a sequence of principles to focus on. Care must be taken not to make the tools or principles the goals. The roof demonstrates that well. The goal isn't TPS, tools, principles or a lofty concept. It is a strong enterprise. The roof boldly promises the "best quality, cost and delivery in the business" for those companies that build this system well. Are you ready to start building?

Throughout this book, I will use the terms "foundation" and "base" synonymously. The far left pillar will be titled "Just in Time" (JIT). The far right pillar will be called "Built-in Quality." The center pillar is titled "Respect for Workers" and is sometimes called the people or culture pillar. The top triangle is the "roof" or results. Toyota guarantees the best quality, cost and delivery, achieved through shortening lead time by eliminating waste.

The point of the house model is simple: just as in building a house, there are a sequence and organization for the TPS principles and tools. The Toyota House model shows how all of the 20 or so simple but effective TPS tools and principles work together to build a strong enterprise.

The Toyoda Family

Just a few Toyoda family members and production leaders collected and framed what we now know as the TPS. Four men in particular are credited with major additions to the system:

- Sakichi Toyoda founded the Toyoda Group in 1902.
- Kiichiro Toyoda, the eldest son of Sakichi, headed the automobile manufacturing operation between 1933 and 1950 and is best known for developing the JIT production concept.
- Eiji Toyoda, cousin and close confidant of Kiichiro, was Managing Director between 1950 and 1981 and Chairman from 1981 to 1994.
- Taiichi Ohno, a production leader from Toyoda operations, was the chief collector of the tools and principles and is best known as the father of the kanban system.[*]

Sakichi founded the Toyoda Automatic Loom Works company. In 1898, Sakichi perfected a textile loom that stopped automatically when a thread broke.[†] This error-proofed (poka-yoke) machine assured quality and minimized waste automatically. It allowed operators to simultaneously manage multiple machines by detecting errors, and thus greatly increased their productivity. In 1910, Sakichi visited the United States for the first time and witnessed the impact automobiles had on America.[‡] Sakichi Toyoda challenged his son Kiichiro to build a Japanese car with Japanese hands.[§] In 1929, Sakichi Toyoda took a bold risk. He sold his automatic loom patents to bankroll an automo-

---

[*] Becker, Ronald M., "Lean Manufacturing and the Toyota Production System," Automotive Manufacturing & Production, June, 2001.

[†] *Time 100*: Aug. 23–30, 1999, vol. 154 N 78, Asians of the Century: Eiji Toyoda, by Ed Reingold..

[‡] Ohno, Taiichi. Toyota Production System: Beyond Large-Scale Production. New York: Productivity Press, 1988, p. 78. Original Japanese Edition: Toyota Seisan Hoshiki, Tokyo, Japan: Diamond, Inc., 1978.

[§] Time 100, Ibid.

bile assembly business and placed his son Kiichiro at the helm. Ford
and General Motors (GM) already had over 90% of the market in
Japan.* They would need to quickly shave years off of Ford and GM's
lead in technology and know-how to be competitive. This despera-
tion to survive seemed to drive the Toyoda family toward relentless
improvements.

Kiichiro's passion for automobiles developed during a year of study
at Henry Ford's famous River Rouge factory in Detroit in 1929.† He
learned Ford's mass production system but needed to adapt it to Japan's
smaller production quantities. He even made multiple models on the
same production line. This was unheard of at the time in the United
States. Since there were few natural resources in Japan to make cars,
he designed systems with no excess inventory and partnered with suppliers to level production volumes. This system became known as JIT
across Toyoda operations.

Kiichiro was a relentless tinkerer. He reverse-engineered Chevrolet
engines early on and in 1935 developed a prototype called the A1
with Eiji, using the Chrysler DeSoto Airflow as a model.‡ In 1936,
Kiichiro wanted to get his new automotive business off to a great start.
By changing the spelling of the family name slightly, it could be written using eight brush strokes—a lucky number in Japan—so Toyota
it was.§ In 1938, Kiichiro instructed Eiji to build a factory in central
Japan. This facility, in what is now called Toyota City, pioneered JIT,
kaizen, continuous improvement and kanban—all essential principles
in the TPS.¶

After World War II, terrible economic times in Japan caused
President Kiichiro Toyoda to reorganize the Toyoda Group. He
insisted that Toyota managers were duty-bound to avoid layoffs—a
predecessor of their *respect for people* principle today. In 1950, when
lenders set terms requiring that Toyota lay off excess manpower,

---

* Becker, Ronald M., Ibid.
† Womack, James P., Daniel T. Jones, and Daniel Roos. The Machine That Changed
the World, New York: Simon & Schuster, 1990.
‡ Dawson, Chester, BusinessWeek, "Kiichiro and Eiji Toyoda: Blazing The Toyota
Way," May 24, 2004.
§ "History of Toyota," http://www.aygo.co.uk/
¶ Dawson, Chester, Ibid.

Kiichiro felt that he must step down as president. Labor unrest ensued. He returned briefly two years later, but a cerebral hemorrhage ended his career.*

Eiji joined Toyoda Automatic Loom Works in 1936 and was named Managing Director of Toyoda Automotive Works a decade later. After visiting Ford Motor Company operations again in the 1950s, Eiji undertook major redesigns of Toyoda automotive plants and even implemented Ford's version of an employee suggestion system. This later became another major building block of TPS—continuous improvement or kaizen.†

Loom machinist Taiichi Ohno was the chief collector of improvement tools. He is considered to be the creator of the TPS as a whole, even though the Toyoda family had many of the pieces in place. Ohno joined Toyoda Automatic Loom Works after graduating from Nagoya Technical High School in 1932. As a tough-nosed production leader, he took Kiichiro Toyoda's JIT concepts even further by focusing all employees on identifying and reducing waste. He found ways to drive inventory to near-zero. Taiichi promoted small batches, product flow and kanban. Ohno's production philosophy called for an extremely skilled workforce. Knowledge flowed freely under his leadership.‡

After a study tour to the United States, Ohno also adapted the simple material supply functions he saw at American grocery stores for auto production. Just as a stock boy used a replenishment card to bring more groceries to the shelf, Ohno used similar cards within his factories and even with local suppliers.

During World War II, Ohno's Loom Works facility was converted to car and truck parts production. World War II took a huge toll on the physical plants. Eiji Toyoda managed the rebuild of all Toyota plants after their destruction. Taiichi Ohno played a major role in establishing JIT principles and what is now known as TPS methodologies. The new facilities were built from the ground up to make components and assemble cars using TPS principles.

* Hino, Satoshi, Inside the Mind of Toyota, New York: Productivity Press, 2006. Original Japanese Edition: Toyota Keiei Shisutemu no Kenkyu, Toyko, Japan: Diamond, Inc., 2002.
† Becker, Ronald M., Ibid.
‡ Dawson, Chester, Ibid.

Growth—The Growing Need for a Documented System or Model

In Taiichi Ohno's early years, very little of TPS was formally documented. Hand-written thoughts and sketches were sometimes used, particularly from Ohno and the Toyota team's mid-1950s visits to Henry Ford's operations as well as supermarkets near Detroit. Toyota workers on the shop floor implemented and perfected the simple but revolutionary principles. There did not seem to be a dire need to formally document the growing body of knowledge as it grew out of the operations led by Taiichi Ohno.

In the tight circle of facilities making up Toyota City in Japan, best practices spread quickly. But, as Toyota spread its production system to its Japanese supply base, and then toward other countries like the United States, a different type of problem grew. How would Toyota communicate their philosophy of production to every worker outside of Toyota City, throughout their supply base, and then throughout the entire world? By 1992, plants produced Toyota vehicles at some 30 sites in more than 20 nations besides Japan.*

Further compounding the need for a documented approach were the positive differences in Japanese culture and the people themselves. Both conformity and acceptance of best practices played a role in the rather easy spread of Toyota principles and practices in Japan. But, Toyota's step-by-step, disciplined, one-best-way, demonstrative training was not always received well by workers outside of Japan.

In the Japanese culture, a saying was often used: *Deru kugi wa utareru*. This means "The nail that sticks up gets hammered down." There is a penchant for conformity in Japan. It is seen as a positive thing to look, talk and even work in like ways. To the typical Japanese worker, being egocentric or overtly unique are negative qualities in such a way that they are not in the West. Add to that a respect for others, and you get more rapid acceptance of best practices or standards. This was not true throughout Toyota's widening supply community.

Taiichi Ohno sought to develop an ideal production system, one that continuously flowed, like a stream. His ideal system was inspired

---

* The Toyota Production System, (Handbook) International Public Affairs Division and Operations Management Consulting Division, Toyota Motor Corporation, 1992.

by Eiji Toyoda's observations at Ford Motor Company in the mid-1950s. Ohno saw workstations very close together, work that was balanced and synchronized, and no inventory between stations. The process delivered finished product to the customer exactly when needed (just-in-time). Taiichi Ohno often asked his peers and supervisors what prevented a no-inventory system. Then, he demanded, "Eliminate the reasons." The result was a production system that eliminates the reasons for inventory.

Taiichi Ohno probably would not have formally documented any part of the TPS on his own. His personal style of prepare-and-demonstrate training was effective in producing disciples that *thought* like him. One of them was Fujio Cho, who would later lead Toyota's American operations through steady expansion in the 1990s. It was Cho who first documented the Toyota principles in a house model.* Cho's model would give the world a simple framework to better comprehend TPS. Hiroyuki Hirano and others would add to the documentation.

The Toyota House model is a good, simple representation of Toyota principles. The strongest analogy behind the house model is that we are actually building *something*. Something physical that will last. Something that is bigger than the sum of its parts. Something that we are proud of. We use tools to build and maintain a home. But, you will not see any tools when you look at a finished house. One negative aspect of a house is that it tends to break down or fall apart and need repair. Thus Toyota desires that all of its employees make their house stronger and hopefully *a little better every day*.

A house is a system of many parts with each part equally important. No one part can accomplish the mission on its own. In fact, they are all tightly linked. It is hard to see where one part ends and the next begins. For example, *robust processes* are not possible without *standardized work*. If you cut the downtime on machines by a lot (*preventive maintenance*), but your *suppliers* do not provide you product consistently on time, you are no further ahead. *Standardized work* and *visual management* serve to highlight *waste*. Each principle and tool is inexorably weaved together into the tapestry of TPS. You can't pull the principles apart, and you can't sustain them separately.

---

* Liker, Jeffrey K., The Toyota Way, New York: McGraw-Hill, 2004.

Not Tools

One of the biggest misunderstandings of TPS is that it is merely a set of techniques or tools in a toolkit. This could not be further from the truth. In fact, each principle or tool seems to serve a singular purpose. Each principle or tool *highlights waste*, which causes a worker to do something to reduce the waste. Shigeo Shingo, professor and Toyota consultant, described it as "system for the absolute elimination of waste."* Just as a good carpenter uses tools to build a home, *process architects* in an organization use the tools to build a strong enterprise.

The chief purpose of every TPS tool is to make waste stand out. Whether it is visual management, 5S, standard work, spaghetti diagrams or Single Minute Exchange of Dies (SMED), these tools serve to highlight waste. The tool itself does not solve the waste problem, but a reasonable person using the tool could be motivated to make a suggestion to reduce the waste. The tool *highlights* waste while the user *reduces* the waste. Sometimes the use of a tool starts a worker's relentless pursuit of waste and reduction strategies.

Here is a good example of how tools highlight waste and cause positive changes. In the line balance chart below (Figure 1.2) collected from direct observations, Worker A is underutilized, while Worker B is overutilized. If Worker A could be cross-trained to do just a few of Worker B's tasks, the entire cell or line could produce more products. The idea to make this change came from the workers after reviewing the line balance chart. A balance chart was just a tool that helped build a stronger process.

Another example demonstrating how tools highlight waste is Figure 1.3, a spaghetti diagram. This diagram highlights the before and after walking waste of the employees working along the production line. Immediately after documenting this waste, the workers came up with simple solutions to reduce it.

Principles versus Tools

There are timeless, unchanging, unwavering principles listed in the foundation and three pillars. These always work in some way, shape

---

* Shingo, Shigeo, A Study of the Toyota Production System, New York: Productivity Press, 1989.

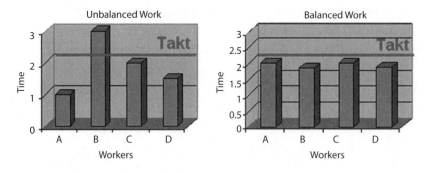

Figure 1.2 Line or Cell Balance Charts (before and after kaizen).

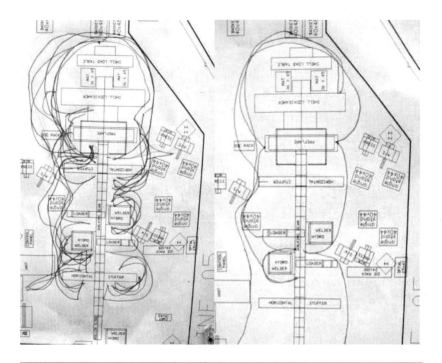

Figure 1.3 Spaghetti Diagrams (operator walking charts; before and after kaizen).

or form. We do not compromise on these principles. And, they are implemented in a certain sequence.

Some of the timeless, unchanging, unwavering principles in the foundation are standardized work, robust product and processes, preventive maintenance, kaizen (small changes for the good, or continuous improvements), (early) supplier involvement, PDCA (Plan-Do-Check-Act) and scientific method, waste elimination and visual management. These will be described in the coming sections.

Key principles in the JIT pillar are right part, right time and right amount, continuous flow, and pull. Key principles in the Built-in-Quality (BIQ) pillar are "never pass bad parts to the next process," make problems visible, and stop-the-line authority (using andons). All of the words in the people pillar (respect, flexible, capable and motivated) are principles with the belief that people are the key to the TPS.

There are also tailorable tools in the house. The tools may look different when applied in different areas, but the principles remain the same. Some tailorable tools shown in our Toyota House are one-piece flow cells, kanban, quick setup, level production and takt time in the JIT pillar. Also, we show problem solving, error proofing, visual controls and person–machine separation (jidoka) in the BIQ pillar. These tools can be used in manufacturing and in other industries, including healthcare systems.

Some additional tools are often used in transforming processes but are not shown specifically in the house, like waste walk forms and 5S with red tag forms in the foundation. In the JIT pillar, many line balance and cell creation tools exist, as do total productive maintenance forms. In the BIQ pillar, tools integrating error proofing into failure modes and effects analyses, and auditing andon responses are inferred but usually not shown. The absence of a tool in this model does not mean they are not used. However, the presence of certain tools in this version of the house shows some significant or important tools. The Toyota House is meant to show mainly the principles and where they are first used.

A good example of principle versus tool is the kanban card. Kanban is a tool. But, the higher principle is called Pull (or Pull systems). There are many ways to pull products (e.g., outlined squares called "kanban squares" on the shop floor, a few dedicated containers, etc.). But, the principle remains: *flow where you can* and then *pull where you must* (and never push). Kanban cards are actually a weak third choice, after one-piece flow and simple, visual kanban.

Lean, TPS and Coming to America

Widespread recognition of TPS as the model production system grew rapidly with the publication in 1990 of *The Machine That Changed the World: The Story of Lean Production* by Dr. James Womack, Daniel Jones and Daniel Roos. This book was the result of five years of research led by the Massachusetts Institute of Technology (MIT). Researchers there found that a certain style of Japanese production management was so much more effective and efficient than traditional, mass production that it represented a completely new paradigm.

This book was the first to coin the term *lean production* to indicate this radically different approach to production. The term was coined by John Krafcik, a research assistant at MIT with the International Motor Vehicle Program in the late 1980s. He worked for GM at the time while seeking an advanced degree at MIT. John is now CEO and President of Hyundai Motor America.

At the same time, other manufacturers were studying more efficient production systems. They developed concepts very close to key parts of TPS but rarely had all of the principles in their models. Some of the companies' systems were called:

- Stockless Production—Hewlett Packard; a great video showing a simulation of these principles was produced on a low budget and copies still exist today
- MAN (Material as Needed)—Harley Davidson
- MIPS (Minimum Inventory Production Systems)—Westinghouse
- Lean Manufacturing/Production—MIT

Throughout this book, I will use the term *TPS* to mean the entire system of production and thinking perfected by Toyota. You will often hear the term *lean* used synonymously, but TPS seems larger than just the lean concepts in *The Machine That Changed the World*. TPS is maintained and improved through iterations of standardized work and kaizen (small, daily continuous improvements), following a PDCA cycle of improvement. The PDCA improvement cycle, from Dr. Walter A. Shewhart at Bell Labs and later Dr. Edwards Deming, uses the scientific method: simplified process steps in the scientific method are to describe a hypothesis, plan a change, experiment to test the hypothesis, check for improvements and then adopt the new standard, if better.

First View of the House

In my desire to learn all things about Toyota and its production system, I took a job at Johnson Controls, Inc. (JCI) in their automotive seating division. My job would be to learn the best methods being used and taught at our Toyota Camry seat-making facility, document them, and then spread the best practices to all 50 or so of our automotive facilities in North America. This assignment put me on a track to learn from the very best, document it carefully, and even get paid for doing it! It was a great job and a great assignment.

My first view of the Toyota House came during an early 1990s meeting with one of my mentors within JCI, Phil Beckwith, and a Toyota *master sensei* for suppliers, whom I will call Hiroyuki Nohba. I asked my sensei, "Please tell me all about the Toyota Production System." Hiro pulled out a blank piece of paper and drew a house similar to the figure shown earlier. The Toyota House. It was simple. Additional tools and principles could be added where they best fit. I made mental notes to learn more about this important learning tool that seemed unique to Toyota. Please read the story "You Will Fail" (in Chapter 2) for more details on this and subsequent meetings. I really had no clue how important the foundation was in these first meetings. I would learn—the hard way!

Hey, My Favorite Tool Is Not in This House

The house model is not meant to be a complete list of tools. To simplify a survey course on lean tools that I was developing for the University of Michigan, I poured the 20 orso unique TPS tools into the house model where I felt they were first used. What was the response? Rather than clarity, the participants seemed to have even more questions about the location, terminology and the use of Japanese words in the house. People were especially vocal about which tools were missing from my version of the house. Some also argued passionately that the variation reduction tools should not appear "under" lean. In other words, the beauty and purpose of the house model was undone by just filling the house with tools. After that, I decided to emphasize the principles and just a few tools on a Toyota House slide closer to the original one drawn for me.

The bigger question is "How will we achieve the principle at hand?" For example, how will we achieve a foundation of stability for our operations? The answer to that question will not include a forced jamming of tools where we think they might fit. In fact, you may not be able to answer that question fully until you have tried several combinations of tools to support the principles. The key to the house model is to deploy and sustain the Toyota way principles.

Sequence

You implement TPS principles and tools approximately in this order: Foundation, then JIT and BIQ pillars together, all the while teaching them to the People that do the work every day and not just some separate Lean or Six Sigma team. The beauty of the house model can be seen in this sequence.

The most common question asked of me in courses at the University of Michigan is something like, "I've done a lot of continuous improvement stuff, but how can we make it 'stick'?" Throughout this book, you will learn not only from my how-to's, but also from my how not-to's. The sequence and principles in the Toyota House can be your guide to keep you from crashing on the rocks as you navigate through the dark waters of transformation.

To build a house, the foundation must come first. Next, you would put up the walls. They need to be erected together. They need to be square and true while complementing each other. If you deploy the JIT pillar principles correctly, you will start moving product to your customers from end-to-end very fast (lead time). If you go very fast, what usually happens? You start making mistakes fast. The house model shows that you must erect the JIT and BIQ pillars at the same time. You need the BIQ tools to maintain speed without errors. And, you need JIT tools to create processes with quick feedback to help make problems visible.

Lastly, you can't attach the roof with a missing wall or pillar. Yet, I have seen many companies either stop short or skip a wall such as BIQ or even Respect for People. They usually have strong competitors because of these missing pillars. Once your three pillars are standing, you should start to see results. In some smaller value streams, customers have seen results within a few weeks. In larger value streams and those with highly complex processes, it may take months and even years to see the impact of the systemic changes contained in the Toyota House model.

The best quality, cost and delivery are possible through everyday application of Toyota principles—that is, to design better processes through shortening lead time by eliminating waste.

What Is a Sensei?

The term *sensei* came into common use through the martial arts but has its origins even earlier in the Far East. Wikipedia says that a sensei

> is a Japanese title used to refer to or address teachers, professors, professionals such as lawyers and doctors, politicians, clergymen, and other figures of authority. The word is also used to show respect to someone who has achieved a certain level of mastery in an art form or some other skill: accomplished novelists, musicians, and artists are addressed with the title in this way; for example, Japanese manga fans refer to manga artist Osamu Tezuka as "Tezuka-sensei."

Sensei is also used frequently to describe the teacher in Japanese martial arts classes such as Aikido, Judo, Jujitsu, Karate, Kendo, Ninjutsu, Iaido and Kenjutsu to name a few.

In the United States, the term *sensei* has also taken on slightly different connotation. It can refer to an outside expert who can provide expert coaching on how to achieve organizational effectiveness. The title *Lean sensei* has since become a common term for describing an expert who can provide advice on operational and organizational strategy.

In my daily role as a TPS coach for organizations, I meet many people who call themselves sensei. In most cases, this could not be further from the truth. The main purpose of a sensei is to keep a client out of the many bear traps along their path toward sustained change and eventually to leave them self-sufficient. This transformation takes years, sometimes even a decade. For the purposes of this book, I will use sensei only to mean a master system-building coach who has deeply developed skills in TPS principles through experience and was also trained by a sensei.

There is no certification for sensei, and I do not believe there should be one. Experience is the real coach. The sensei merely guides the student through a series of learning experiments. The sensei does not prevent the student from experiencing all of the problems along the way. In fact, part of a student's learning might be to hit a roadblock or two and then reason his or her way out. The sensei will, however, prevent the learner from serious mishaps and risks by firmly reigning in the student—much like Mr. Miyagi prevented Daniel-san from running too far ahead of his training in the movie *The Karate Kid*.

Goals of TPS

One critical conclusion needs to be drawn before discussing the details of the Foundation and three pillars. The goal is *not* lean. The goal is *not* to implement lean tools or principles. The goal is to *build* a *strong enterprise*. The Roof of the house states "Highest Quality, Lowest Cost and Best Delivery." Toyota made a fairly iron-clad guarantee to the leaders at Johnson Controls. They said, "*If* you deploy and sustain the principles of the house, you will earn the best Quality, Cost and Delivery in the business—guaranteed!" Our sensei from Toyota was right.

Some people note the lack of their particular "tool" in the house shows ignorance of continuous improvement. I don't see it that way. Understanding the details of this model gives you a golden roadmap for building and sustaining a strong enterprise. There are many more details behind the painted walls of this house. That is why a house shape is a great model for implementation.

Changing Culture

Don't wait for a culture change before you begin your TPS transformation. Culture change comes through handing workers the keys to their process just like when you hand over keys to their own car. As people are given the keys to change their own work processes, their behaviors will need to change as well. If associates change their behavior, the culture will change. This is why we say implementing TPS gives you "buy-in through involvement." Toyota does not buy training to fix their culture. They just improve their work processes, every day, without end.

←---- Active, Visible Leadership ----------------------------→
Workers change processes → Behavior changes → Culture changes

All of these changes are guided and encouraged by leaders who trust and respect their workers. Leaders set and clarify the goals. Leaders teach and model the TPS principles. Workers identify and remove waste from their processes daily. This process is repeated day after day. Once employees see that no one is wrongly removed from employment, and their ideas are respected, a visible culture shift occurs. You can see it in the attendance sheets. You can hear it in the hallways. And, you can see it in the shared celebrations and rewards.

2

FOUNDATION OR BASE PRINCIPLES AND STORIES

Foundation

Foundation of Operational *Stability*		
Standardized Work		Robust Products and Processes
Preventive Maintenance	Kaizen	Early Supplier Involvement
PDCA-Scientific Method	Waste Elimination	Visual Management

If you wanted to build a house or structure that will last, you would make sure that your foundation was solid, square and level before building the sides. Wouldn't you? As I consulted for literally hundreds of organizations, one common thread stuck out. Companies diminish the importance of deploying and sustaining "base" principles. When their transformation efforts stumble, they assume that the Toyota principles do not work. As Nohba-san once told me, "It's not the Toyota principles that failed, it is you. Don't skip the foundation!"

Some of the key principles in the foundation are:

- Standardized work—There is one best way to do work (documented).
- Preventive maintenance—All critical equipment and resources (including people) must be up and available to you a high percentage of the time (e.g., 99%) before attempting advanced principles like one-piece flow cells.

- Robust products and processes—All processes are repeatable and consistently produce high quality parts; all products are well designed and easy to build.
- (Early) supplier involvement—Involve suppliers of critical components early; use their input and ideas to drive down cost and lead time.
- Plan-Do-Check-Act (PDCA)/Scientific method—Daily, team-based problem solving should use PDCA thinking steps and the scientific method; find it—fix it.
- Waste elimination—Every team member is taught what waste is and is given authority to creatively reduce waste; also called developing "eyes for waste" in all associates.
- Visual management—Make every workplace so visual that anyone can walk in and visually understand the current situation.

Just as the foundation comes first in a house, these principles must take root in order to pave the way for the walls or pillars. All of these principles are continuously improved. But, you need to get a good start on each of them before proceeding to more advanced concepts. There is a sequence to the house. If you start skipping principles because they are hard, your house will be very difficult to build and will be unstable.

Standardized Work

First and foremost in this foundation is **Standardized Work**. This principle states that there is ONE best way to do a task. It's the best way that is documented, today. We may improve it tomorrow. It is this staunch commitment to the best way, while encouraging improvement that balances Toyota's Standard Work system. Another worker's new method may be "better" if it is safer, higher quality, faster and lower cost, in that order.

There are many ways to document the best method for a task. The format does not matter as much as the worker's involvement in developing the best method. How are you doing on this principle? It's only the first.

You may be asking, "Will we *ever* be done with continuously updated tools like Standardized Work?" Good point. You can't. But, you do need a good start on all of these principles in the area you will transform first. Nothing in lean is subjective. Since everything is clear and objective, let's define "good start" as a level of implementation where you can tell people in your company, "Come and see. This is what I mean by a Standardized Work system." If your associates can see and understand the changes, you have a good start.

Please read my stories "You Will Fail" and "Timeless, Unchanging Principles" for humorous views of how I skipped all base principles at Johnson Controls, Inc. (JCI).

Preventive Maintenance

The next principle in the base is **Preventive Maintenance**. This principle requires basic blocking-and-tackling on your key equipment and resources by your skilled maintenance professionals and people that know the equipment well. They should not break down at unplanned times. A related tool is Total Productive Maintenance (TPM). We will cover TPM in more detail in the Just-in-Time (JIT) pillar. In TPM, the operators of the equipment volunteer to do basic cleaning and maintenance to take care of their equipment. This is usually quite effective in reducing unplanned downtime, since the operators are around the equipment more often and know it well.

Something that does not come out as clearly in the Toyota House is how well each of these principles applies *above* the shop floor or even in complex healthcare processes. Sometimes the name needs to be altered a bit. My sensei at Triangle Kogyo, whom I'll call Bo Shimono, once told me, "If you think the principles apply well on the manufacturing floor, wait until you see them in administrative processes!" He was right.

For example, in most non-manufacturing processes, the key "machines" are people. Since the principle behind preventive maintenance is having all of your key machines up and available to you when the cell is running, it follows that Preventive Maintenance for administration and knowledge-worker processes would be having flexible, capable, highly motivated people available and cross-trained when you need them. How well did you handle vacation periods last year? Was all of your work done for you by the time you returned? Probably not. We have a long way to go before we check the Preventive Maintenance box as complete.

Kaizen

The next principle in the base is **Kaizen**—daily, incremental improvements. The direct translation is close to "small change for the good." The Toyota Production System (TPS) is powered and improved through iterations of Standardized Work followed by kaizen, followed by standardizing work again, and repeated. Every employee is responsible for bringing forward and implementing ideas for improvement every day! It is this relentless pursuit of kaizen, of eliminating waste, daily that drives TPS.

Robust Products and Processes

Robust Products and Processes refers to the consistency and repeatability of the output of the process. Process quality, capability, reliability and repeatability are wrapped up in this simple phrase. How are you doing on that one? Does your process produce the same output if different staff work them? If you don't have a handle on this principle, please don't slap together machines into a one-piece flow cell! You'll just make garbage quickly.

Early Supplier Involvement

Early Supplier Involvement is a huge principle in TPS, but not much is written on the subject. In this principle, suppliers are treated as partners and are also trained in the TPS methods. They are treated as family members, not as slimy contractors. Toyota also understands that all unnecessary cost and waste are eventually borne by Toyota or the customer. This is a key reason why Toyota wants their partners to understand and use their principles in production as well.

Not all suppliers are single-sourced and brought into the new product development process early. But, Toyota and others would do this for suppliers of strategic components. Many experts, including David Nelson of Honda, suggest different *strategies* for different components based on their technical challenges (or risks) and cost.* For high

---

* Nelson, David, Patricia E. Moody, and Jonathan Stegner, The Purchasing Machine, New York: The Free Press, 2001.

part-count assemblies (like automobiles or even refrigerators), supply chain experts recommend a strategy of quadranting the components into high and low **technical content** (or sometimes risks to your company) along one row, and then into high and low **cost**. It is possible to use a *lowest bid* strategy on some low-tech, low-cost items. But, an assembler should never risk his or her reputation on this strategy for components with high technical or engineered features. Toyota and others would partner with a single supplier or divide their product lines to just two suppliers for these strategic-quadrant components.

Toyota partners with a higher number of long-term suppliers than most other assemblers. These suppliers give a lot more of their talent and ideas to Toyota for many reasons. Two key reasons are trust and willingness to share technology secrets with Original Equipment Manufacturers (OEMs).

Dr. John Henke, professor and president of Planning Perspectives, offers a glimpse into this principle in his annual OEM-Supplier Working Relations Index.* Toyota and Honda are putting larger gaps between themselves and the competition in the United States. Here is a single point of proof. I was once on a supplier team that moved from a Big Three OEM project to a Toyota project overnight because the Big Three actually showed our designs and innovations to our competitors, who took them and underbid us. We recouped none of our design costs. This smart supplier moved their top talent and engineers to Toyota and Honda programs. At least they didn't give away our innovations to our competitors! We assigned the interns and plodders to the Big Three programs. It only took a few of these learning exercises for a strategic decision to be made—no innovations or superb engineers for Big Three programs.

I added three tools to the usual Toyota foundation: PDCA-Scientific Method, Waste Elimination and Visual Management. Even though these are not elevated to principle status in the original Toyota House, I find they meet the definition. Webster defines a principle as a "comprehensive and fundamental law, doctrine, or assumption; a rule or

---

* Reuters (www.Reuters.com), Annual Supplier Rankings of Automakers by Suppliers Show Toyota and Nissan Slipping, Ford Gaining and Chrysler Tanking (Press Release), Aug. 11, 2008.

code of conduct; or a habitual devotion to right principles (e.g., a man of principle)."*

Solidly in the base is root-cause PDCA problem solving. Problems are not to be ignored or covered up. They need to be solved—today! A person could put a bandage on a deep cut, but only sutures will allow proper healing. The Toyota way of improvement uses the tried-and-true quality improvement **PDCA** thinking steps and the **Scientific Method**. A hypothesis is set up (e.g., if we make our work area highly visible, it will take less time to choose and match up parts, and will cause fewer errors). If you can test the hypothesis at low or no cost, work is treated as an experiment. The quick experiment either proves or disproves our hypothesis. This is why proper process-level metrics are so important. We are trying to fly by the numbers of a few key metrics. We will discuss metrics more when we discuss the "roof."

Every employee must be trained to have "**Eyes for Waste.**" Their jobs are not in jeopardy. The enemy is waste. A work task is considered waste or non-value-added if it takes up time and costs us money but does not add any value in the eyes of the customer. If you have 500 employees at your site, you need to develop 500 waste-seeking missiles. When they see waste, they "blow up" with a suggestion to reduce the waste.

Taiichi Ohno categorized waste into seven types. Nearly all waste can fit somewhere in these categories. If we add another waste, "Not utilizing the talent and ideas of our workers," we can spell an easy to remember acronym—DOWNTIME. Downtime is bad. Waste is bad.

The categories of waste with brief descriptions are shown below:

1. Defective and rework—Defective products impede flow and lead to wasteful handling, time and rework.
2. Overproduction—Producing *more* material than is needed *before* it is needed is the fundamental waste in lean manufacturing; material stops flowing.
   a. Problem: The overproducing station is making something that is not needed today *instead of* something that is needed.

---

* *Merriam-Webster Online Dictionary*, http://www.merriam-webster.com

   b. Problem: The overproducing station is making more prod-
      ucts that are not going to ship today *instead of* walking
      down to the current bottleneck and helping.
   c. Problem: The excess inventory covers up other problems
      and wastes.
3. Waiting—Material waiting is not flowing through value-
   added operations; people, equipment and material can be
   waiting.
4. Not utilizing the talent and ideas of our workers. (New)
5. Transportation—Movement of goods and materials does not
   enhance the value of the product to the customer.
6. Inventory—Material sits, taking up space, costing money and
   potentially being damaged; problems are not visible.
7. Motion—Worker motion and movement do not add value to
   the product.
8. Extra Processing—Extra processing not essential to value-
   added from the customer point of view is waste.

A simple way to uncover waste in your operations is to do a "Waste
Walk." This exercise takes only about 20 to 30 minutes. Hand your
workers a copy of the wastes in the previous list. Ask them, "Do you
think you can find any waste in our operations?" They will surely say yes.
Tell them, "You have 20 minutes to find one specific example of each
category of waste. Then, write down next to it one suggestion to reduce
that waste." You will be shocked how many great ideas this waste walk
generates. But, make sure the workers put their names on the form. If
the suggestion would require low or no cost to deploy and will not hurt
safety or quality, circle it and hand it back to the employee for them to
quickly implemt.
   Developing Eyes for Waste is one of the most critical parts of your
TPS journey. To conclude this brief section on waste, I will let Taiichi
Ohno's words do the talking:

> Our system is so far from generally accepted ideas (common sense) that
> if you do it only half way it can actually make things worse. If you are
> going to do TPS you must do it all the way. You also need to change the
> way you think. You also need to change how you look at things.

Just as magicians have their tricks, the gemba technique has its tricks. The magician's trick in this case is "the relentless elimination of waste". In order to eliminate waste, you must develop eyes to see waste, and think of how you can eliminate the wastes you see. And we must repeat this process. Forever and ever, neither tiring nor ceasing.*

In **Visual Management**, we attempt to make a workplace so visual that anyone can walk into a workplace and visually understand the current situation. For example, an inventory control manager could keep counting and reminding the workers to stop producing when finished goods inventory reaches a certain level. Or, leaders could paint outlines around a small area near final assembly with a line on the back wall stating "stack only this high!" Visual management beats nagging every day. But, leaders still need to walk the floor and audit that these visual controls are being used every day with discipline. Please read the story "Stand in Circle" for more info on Visual Management.

My Most Interesting "Foundation" Stories

The stories that follow will give additional insight into the **Foundation** tools. Humor is a great way to learn something. Please read these and reflect on the questions that follow each one.

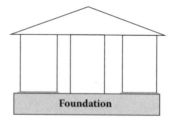

(University) Time Study and Sharp Objects (1)

"We will win and you will lose. You cannot do anything because your failure is an internal disease. Your companies are based on Taylor's principles. Worse, your heads are Taylorized too. You firmly believe that

---

* Ohno, Taiichi, Toyota Style Production System—The Toyota Method, Toyota Education Department, Jan. 1973; from the foreword titled "Practice, Not Theory."

sound management means executives on the one side and workers on the other, on the one side men who think and on the other side men who only work."

Konusuke Matsushita

As part of my Industrial Engineering education, we learned how to do formal time studies. In class, we watched boring videotapes, applied the technique, and thanked God for the ability to rewind the video as we missed several key tasks. Eventually, we muddled through well enough to fool the professor. But, could we do time studies on *live* operations?

Our Work Measurement class professor worked full-time at a downtown hospital. He was involved in all kinds of special projects, and showed us how Industrial Engineers could apply our tools in a healthcare setting. Our professor told us that we would practice live time studies on nurses. Cool.

We showed up in the lobby and walked up to one of the floors. We received our watches and some brief instructions. We were each pointed to a nearby nurse and told, "Go get them." I observed a wary nurse from a distance for a while, then wrote down what I thought were her process tasks. I pulled out the stopwatch and started to click the lap button while I hectically recorded times. She frowned. I was not very good at this. The nurse moved fast. I chased her. As I was nearing the end of my study, I noticed that she had some sharp objects in her hand. She was smiling now. I ran. She chased me.

Our professor talked several of my peers into working at the hospital as Management Engineers. The hospital paid their Co-op students about 20% less, so I joined a large automobile assembler, Admiral Engines, to "change the world." Nurses were scary.

- Why do you think the nurse was "wary"?
- What could I have done different?
- If I told you that about 5% of the nurses that were "time studied" at this hospital lost their jobs after each study, would that change your mind?

(continued on next page)

- If we used our stopwatches only while the nurse was waiting or searching, with the purpose of reducing the non-value-added time, would the results have been different?
- Who was using the tools of improvements (stopwatch)?

(Admiral Engines) Murphy's Law (2)

"Put a good person in a bad system and the bad system wins, no contest."

W. Edwards Deming

My first real job as a budding, young industrial engineer was for Admiral Engines (AE). All industrial engineers needed to serve a stint as a production supervisor. I said, "Let me supervise your worst factory." So, I found myself in Detroit reporting for duty at a transmission assembly facility. The Human Resources (HR) department took up most of my first day with boring presentations intertwined with lots of waiting. Late in the afternoon, HR walked me through the cavernous facility to my new department—the Final Assembly Line. We soon caught up with Sandy, the supervisor who would train me. She was very pregnant, but moved fast. She stopped for just a moment and said, "I don't have time today to teach another snot-nosed college punk!" She said to meet her at 6 am the following morning. I liked her.

Well, Murphy and his law worked. Sandy had her baby six weeks early that night. No one met me at 6 am. In fact, I did not know even one single person's name at AE, except for Sandy and the HR guy, who would arrive some three hours later. I found out just how hard it is to manage production when, for years, workers have been told, "Just do as you're told!" I didn't even know where the restrooms were! At 6:00:01 am my assembly line started moving. At 6:00:02 am my assembly line stopped.

What would you do? They didn't teach me anything in engineering school about how to manage people or assembly lines. I remembered something my father taught me, "When in doubt, ask." So, I introduced myself as the new supervisor and asked the kind-looking woman at the first station what was happening. She smiled and said, "You can't run the line without workers." That made sense. (Observe) Only about half of the workers were at their assigned stations.

I asked, "What do I do?"
She said, "Go get one of your Relief Workers."
Ah ha, progress. I asked, "And who might they be?"
She pointed to a gentleman over by a brick wall that seemed to be painting a beautiful mural—à la graffiti (Detroit style).

I wandered over, as many eyes locked in on me. I tapped the man on the shoulder and started to ask him to help relieve one of our still-absent workers. He whirled around threatening me with a long butterfly knife and said, "Don't you *ever* touch me." Confused, I backed out of the way. My relief worker put the blade back in his pocket and continued painting. Ashen-faced, I walked back to my supervisor stand. After several large men walked briskly through the factory screaming and cajoling, the line started up around 6:28 am that morning. During a brief lunch break, I sat with another young engineer, who was supervising a different line for a few weeks. I told him of my harrowing experience.

He said, "That's nothing. On my first day, someone shot at me in the parking lot!
I asked incredulously, "What did you do?"
He said, "I ducked!"

So went my first day as *the boss* at AE. I realized very quickly that "the inmates ran the prison" here. But after nine long months, I finally figured out what it took to be a good supervisor. The same day I figured that out, AE promoted me to the Industrial Engineering Department! That is another story.

Post-script: Upon reporting the incident to my leaders, a back-office meeting was held where I was asked to apologize to the worker for tapping him on the shoulder. The worker swore he did not have a knife in his front left pocket. Seriously.

- What was your first day like?
- What should a good supervisor do?
- Should support resources like HR, Information Technology, Training and Quality work the same shifts as production? Why or why not?
- Is there a management style somewhere between dictatorship and "inmates running the prison?"

Andre the Giant and Stopping the Line (3)

"Find it; Fix it!"

Taiichi Ohno

Every team has its key, hard-working player. The early Lakers had Magic, my beloved Detroit Red Wings had Stevie Yzerman, and the Broncos had Elway. At the AE Transmission Plant, the Final Assembly Line had Andre. I ran, or more accurately, watched the Final Assembly Line. Andre ran the 12-Gun.

The 12-Gun operation took strength and dexterity. My friend Andre had both. Here is how it worked. After completing the guts of the assembly, the large cap of the transmission was placed on the housing. Andre deftly grabbed (exactly) 12 long bolts from a bin and inserted them into the cap holes. Andre pulled and maneuvered the hanging monstrosity with 12 spinning spindles onto the heads of all 12 bolts. Then the machine perfectly torqued them down—at least in theory. Andre was so good and so fast on the 12-Gun that he had completed his master's degree in history, doing most of his reading between the time he finished his cycle until the next transmission arrived. I didn't mind. He was good. I started and ended each long shift with my usual "Andre!" salutation.

Downtime—the curse of all production! At any point during my tenure as supervisor, only three or four of the 12 bolt-head spindles actually functioned! To make up for the other eight or nine bolts that were not torqued down, I would need to assign a relief worker to the end of the line all day to torque the other bolts down by hand. It was not a good thing.

One day in frustration, I hit the red E-stop button next to the 12-Gun and told my line workers to go to lunch 40 minutes early and stay through their normal lunch time. I didn't have to tell them twice. They were gone before the words echoed off my boss' bald head as he roared toward me. I would not take my hand off the E-stop until he got Maintenance to fix the 12-Gun properly.

It worked. For exactly four blessed days, all 12 spindles worked. Upon returning from their long lunch, my workers looked at me with a bit of admiration, and I had earned Andre's respect. He said, "You got my back; I got yours, boss!"

For the first time in my life, someone had called me boss. I must admit, I stopped that line more out of frustration than cunning. Fact was, I was hoping they would fire me. But, AE had too many other bad apples ahead of me in that category, so they promoted me.

- What takes longer: having the team stop and eliminate the root cause of a problem today, or working around it every day for years?
- Was sending the workers to lunch early the right thing to do?
- If you were the Supervisor, what would you do? How about if you were the Operations or Plant Manager? Why might your answer be different for these two levels of leadership?
- What would a fully trained Toyota supervisor or leader do?

Time Study—Part 1 (4)

"Where there is no Standard there can be no Kaizen."

Taiichi Ohno

After several months as supervisor at AE, I finally figured out my job. You see, I thought I was the boss. Management told me that I was the boss and needed to "stay on them workers." But, if I fetched them parts, kept their machines working, ran interference for them (from management and those pesky engineers), and basically served them, the darn line ran fairly well. The day I figured this out, AE promoted me to the Industrial Engineering (IE) Department. I guess they didn't want any poisonous thoughts like supervisor-as-servant infecting their operations. So, I walked from the smells and sounds of assembly to the small IE Department at the back of the factory.

As a degreed IE, I was excited. Finally, I would get to use all of that stuff my professors were blabbing about. The guy from HR introduced me to the team.

One of them asked, "What should we have the greenie do today?"

The others, feet on desk, decided, "Let's have him do a time study."

Hot dog! I was the best stopwatcher in my IE class when we practiced on videotaped operations.

I said, "Let me at it!"

Warily, I asked, "Are there any nurses here?"

They looked confused.

The head IE reached into a desk drawer, blew the dust off a large clipboard, and handed it to me. I was in awe. I caressed the board as it slid effortlessly into the crook of my arm. The clipboard looked like Picasso's painter's palette. The board was curved and cut out around my elbow. The fingers of my left hand stuck through another hole cut in the top of the board for ease of grip. My fingers touched the top button of a shiny silver stopwatch attached to the board. The top right corner of the board even had a calculator attached! There was nothing this board and I couldn't do!

The head IE handed me a Time Study Form all filled out with the previous motion study tasks and even some standard times.

I asked, "Where do I go?"

Head IE said, "4-banger Final."

I said with excitement building, "That's my old line!"

I asked, "And which station will benefit from my IE prowess?"

They laughed and said, "12-Gun."

I said, "Andre! He's my best worker!"

I ignored the old guys laughing behind me as I held my Picasso board high and nearly floated out to my old line. Wait until they see me now!

Andre saw me out of the corner of his eye, and said, "Hey boss, what are you up to?"

I said, "I am an IE now. I'm going to do a *time study*."

I noticed a twitch in his eye as he beheld my beautiful time study board.

He said, "On who?"

I said, "On you, sucka."

Andre laughed out loud. That wasn't the reaction I was hoping for.

Fear me—I am management now! Look at my time study board and tremble.

My friend laughed again and said, "You're not going to like this!"

I asked, "What do we do?" [Remember, when you don't know...]

Andre said, "We need to wait for the others to arrive."

We shared small talk for another 30 minutes. He worked deftly as we spoke and waited. He was re-reading *War and Peace* between cycles.

The first to arrive at the scheduled time study was the union steward. I knew him—everyone knew him. He enforced one side of the contract very well and got people like the guy with the knife back on their jobs. My old boss, the Area Manager, was the next to arrive. We waited for the last person—the International Union time study expert. He sauntered up swinging a gold watch—a gold watch! I was envious of his gold watch. But, I had *The Board*. My old boss asked if I was ready, then someone said loudly, "Let the time study begin."

Andre...went...just...like...this. He slowed to somewhat less than a snail's pace.

The union reps and my boss turned their backs.

The International expert never even started his gold watch.

I dropped my beautiful clipboard! I yelled, "Hey! He's not working!"

My old boss yelled back, "Just do your job Steve!"

Andre picked up one bolt—just one lousy bolt and slowly turned toward the transmission. It was then that I noticed that the stupid engineering standard actually said to pick up one bolt at a time. Dang! Our worker had learned a better method, but it was never documented and shared. Whose job was that! Three transmissions passed by with no caps, only to be turned over at the next station, where several internal parts fell to the ground.

Line workers yelled.

Mass hysteria.

Don't laugh—you might have bought the car with that transmission!

I tried to focus on the time study and Andre's slow-motion dance. I was to take 10 time cycles for the 12 or so detailed process tasks that Andre was doing. Easy, right? What should have taken me 20 minutes maximum expanded into over two hours of gamesmanship. Andre was enjoying this way too much. After five full cycles of time observations, I gave up with the stopwatch. I turned my back to calculate the average times for each task and fill out the rest of the form. Someone announced, "time study over." Andre went back to his steady, fast method. The chaos and yelling downstream died down.

There it was at the bottom of the Form. My last step was to estimate Andre's *pace*. Pace is an estimate of the percentage of maximum speed that the time study experts felt the worker was performing during the time study.

Mr. Gold Watch said, "Steeeeeve, what was his pace?"

I blurted, "Pace! I didn't need a stopwatch for this. I could have used a calendar!"

He said, "Be serious, Steve. Time study is a serious thing. Now, what was his pace?"

I said, "Can I use a negative number?"

He said, "Be serious Steve!"

I said, "I watched Andre for nine months. He was my best worker. That was only 20% of his usual pace."

The other management person agreed.

But, the two union leaders and Andre all said in unison, "No. That was 100%."

So, a *compromise* pace of 75% was placed on the form without argument. Note: The 12-Gun was actually listed as the bottleneck, if performed one bolt at a time. Because I had the gall to do a time study on the 12-Gun that day, AE was required to slow down the line! I vowed that I would **never** do a time study again! That vow would be challenged later by my first real sensei, Bo Shimono. But that is a different story.

- Who owned the tools of improvement at AE?
- What do you believe was management's purpose for the time study?
- What was the union's purpose during the time study?
- Why was Andre allowed to do what he did?
- What would you have done?
- Do your workers sometimes find better ways of doing things? Do you have a system for respectfully learning, documenting and spreading these improved methods?

(AE Technical Center) The Big Lie (6)

"…automation applied to an inefficient operation will magnify the inefficiency."

Bill Gates

"Work expands so as to fill the time available for its completion."

C. Northcote Parkinson, 1958

"Everything expands to fill all available time, space, etc. (Just look at your garage.) So, don't allow it."

Steve

After working in the IE and Reliability Engineering Departments at the transmission factory, I was promoted all the way up to AE's Technical Center. They didn't know what to do with me, so they put me on the Future Factory (FF) team. I was trained by Eli Goldratt, author of the groundbreaking book *The Goal*, on his scheduling package called OPT (Optimum Production Timetable). They called me a *Jonah*, some sort of theory of constraints expert. I was also learning a lot about computers—like how to force a "message" to another user on another IBM computer using the status bar—a classic nerd's version of instant messaging. With 19 or so other FF'ers, I bought into *The Lie*. *The Lie* was that we could achieve great flow and performance *IF* we created and executed *perfect* schedules. We thought we could trick those "lazy" workers into working harder, if we could just deliver perfect schedules to them.

Because of my experience running a transmission line, they gave me a key role in scheduling machining lines (back shops) for a nearby factory. My job was to find out what went wrong in the factory and bring those changes back to the other OPT analysts, who would then reschedule the operations. Then, I would deliver the new schedules to the supervisors. At 6 am the schedule was perfect. Then people started arriving! Some parts we thought were good were actually bad; some machines went down; some parts were lost (or hidden by the other shift workers so only they would have them later); some people did not show up; and, a few people showed up drunk. It was a typical morning at an AE plant!

Small problem. I could not collect all of the mishaps, run to the computer jockeys, re-run the schedule, print the schedules, and then run back to the floor supervisors fast enough to keep up with the changes that happened moment-by-moment. We tried all kinds of Computer Integrated Manufacturing (CIM) tricks. But, we still could not detect the problems and get the changes into the computer fast enough. We bought a supercomputer company, a water-cooling equipment company (to cool off our overtaxed computers), an artificial intelligence software company, and even a few large-headed PhD's. No amount of money could get us around the *re*-schedule problem. We were stumped.

AE established a joint-venture with Toyota in what I will call Great United Motor Builders, Inc. (GUMBI). The GUMBI experiment results (see People Pillar principles and stories) were reported later that year, analyzed from afar and dismissed as hocus-pocus by most of AE's leaders. But, the FF scheduling team had a different hypothesis. We were certain that Toyota figured out the key to re-scheduling! We begged our boss to send all 20 of us to California to see how Toyota managers at our joint venture GUMBI factory scheduled. But, the mid-1980s were austere times. We were repeatedly denied.

We remained stumped. That is, until Toyota did us a huge favor. They built a factory in Georgetown, Kentucky. We worked just four hours north of there! So, we rented one of those special little busses for all 20 of us, sang our nerdy songs all the way to Kentucky, and anticipated cracking the "code" for Toyota's mystical scheduling tricks.

The bus pulled up. We nearly ran into the new factory. All 19 of my teammates went straight to the computer room—that was their job. I ran straight into the bowels of the factory to discover how they scheduled and re-scheduled. With my escort trailing behind, I charged up to the first worker I could see standing at a machine and blurted, "Where's your schedule?" He seemed startled and said nothing.

I thought to myself (to my southern friends, please forgive me for this; I'm from the north), 'Maybe I am talking too fast.'

So, I put on my best southern drawl and said slowly, "Wheeeere's yoooour sched-yoool?" while outlining the shape of a paper schedule in the air with my fingers.

He smiled and said, "I dooooon't knooow" while also outlining the shape of a paper schedule with his fingers.

Hmmm. They didn't teach me in engineering school how to communicate, get along in teams, do root-cause problem solving, or to do anything useful for that matter. But, I was young and desperate to show my team that I could figure out how Toyota schedules production.

I explained to the gentleman at the machine that my bus was leaving at 3 pm and that I needed to learn Toyota's scheduling system.

I asked him slowly, "Will you make any parts today?"

He said, "Yep."

I said, "Great, which ones next?"

He said, "No clue."

Dang.

At this point I drew a quick conclusion (again, please forgive my youthful ignorance). Not only was this person dumb—he didn't know what a schedule was. He was also lazy—he wasn't even trying to *look* busy. He was just standing there.

I tried a different tack, "Are you going to do *any* work today?"

He said, "Yep."

I said, "Great, what parts will you work on?"

He replied again, "No clue."

As he smiled (and I fretted), I heard something rolling down a plastic PVC pipe pointing at the aisle way. Roll, roll, roll, thunk. He picked up a green painted golf ball and said, "Now I'll make the green ones." Here I was, an IE/OR, Factory of the Future team member, OPT analyst, Jonah trainee, theory of constraints expert. It was going through my mind like, "He will make the greeeeen ones"? He was clearly enjoying the puzzled look on my face.

You see, this gentleman was actually a Quality Circle leader and master-degreed ex-school teacher. Most of you realized by now that he was just pulling my leg. It took him just five minutes to undo almost five years of indoctrination by AE and my engineering professors.

He said, "We have a schedule—the shipping schedule. All other operations wait for their *pull* signal."

I said, "What is pull?"

He said, "Our assembly line flows cars out the door according to the shipping schedule. Subassembly operations like mine are linked to the main line by a small quantity of inventory. *Pull* is simply a signal telling us to *replenish* what has just been used."

Replenish what has just been used...

The light bulb went off in my head.

I asked, "So, if the machine ahead of you goes down..."

He finished, "Right, I wait for a signal from another machine that is still running, or go help elsewhere."

I noted how this pull signal automatically re-scheduled his operation.

I said, "And, if the luxury model or other options start selling more..."

He again finished, "Right, I'll set up and make the green ones."

Wow! This was too simple. Not even I could mess this up. Visual factory re-scheduling with something called *pull signals*.

The worker said, "No one makes anything until we get a pull signal. If you would have guessed, or I would have guessed, or a scheduler would have guessed, we would have said the red high-volume ones would be needed next. We would have been wrong! I would have been through with my quick setup and making the wrong parts. Then, the assembly operation ahead of me would run out of parts. Everyone on that line would be sitting on their hands because of our wrong guess."

That happened most of the time when we scheduled, or guessed. He demanded, "You follow this container to the assembly line. My customer will take the last part out of a similar container, and then my container will be shuffled in JIT. If we would have even started making something else, we would shut them down." He said, "Follow some other containers around the factory. Don't ask the workers for their schedule. They'll think you're a moron! Ask them what their *pull signal* is."

I did just that. I followed his container to the assembly line. Sure enough. The container arrived just before the assembler's container ran out—JIT. Not even one minute to spare. I saw other pull signals. Sometimes it was an empty container, sometimes it was an outlined spot on the floor, sometimes it was a plastic laminated (kanban) card, and sometimes it was a well-labeled hand cart. No one started making anything until they received *permission* to produce it. This was so different from and so much better than AE's elusive "magic schedule." By the time we printed each schedule, it was already bad. I had to tell my friends.

I searched the entire Toyota factory. They were nowhere to be found. My escort brought me back to the computer room. All 19 of them were still there, pouring over ream after ream of printed lines of code and output from Toyota's Material Requirements Planning (MRP) system. The poor Japanese host kept shrugging his shoulders as my teammates peppered him with questions. Toyota was using MRP to get raw materials into the factory on time, but they sure were not using it to schedule their operations!

I better let them know...

I cried out, "Eureka, I have discovered the key to the scheduling problem!"

My bleary-eyed teammates looked up, hopeful.

I said, "The key is, don't schedule."

They scowled and went right back to their reams of code.

I was dumbfounded. I tried to tell them of flowing assembly lines and simple pull signals. Just after 3 pm, we re-boarded the special bus. It was not that I was so smart and the others were dumb. I happened to be the youngest person on that bus. But, the rest of my team concluded, "It will never work. They'll never make cars in Kentucky!" Pride goes before a fall. I sank deep into my seat as my teammates trumpeted the superiority of AE's systems and prowess.

Toyota made over $17 billion in fiscal year ending March 2008[*]! I *had* to learn more about TPS. By the time my bus reached the Michigan border, I had decided to quit my job at AE and focus on learning more about the Toyota way of doing things.

---

[*] www.hoovers.com

- What are the reactions of your leaders if they are shown data demonstrating better performance at a competitor?
- What is better one-piece flow, Pull or Push (scheduled operations)? Why is Flow better than Pull?
- Would you have quit working at AE?
- How much do you learn each day at work? You can learn a great deal (how not-to) even from weak companies.

(Glass Company) Five Why's (7)

"A relentless barrage of *why's* is the best way to prepare your mind to pierce the clouded veil of thinking caused by the status quo. Use it often."

Shigeo Shingo

As I searched for a way to learn from Toyota, I moved on to a supplier to AE and helped them make windshields and other glass products. I became somewhat of an expert in discrete event simulation modeling. We simulated every new manufacturing process before we re-built them from the ground up. My engineering teams were sent to each factory with orders, "Do not come home until the factory is performing at your design targets for production and quality." We spent many long months before someone mercifully sent us back home, usually far from those *ideal* targets.

Around that time, a great book describing parts of TPS came out. It was called *The New Manufacturing Challenge* by Kiyoshi Suzaki.* The book was red, so my friends who were also pursuing TPS and I called it *The Red Book*. It is still one of the best books written about waste-ology. His caricatures of sweating workers are carved deeply in my mind and still make me laugh. Kiyo was describing every factory I ever worked in, and it wasn't pretty. I devoured this book. Then, I read it again.

---

* Suzaki, Kiyoshi, The New Manufacturing Challenge, New York: The Free Press, 1987.

One important concept he promoted was using the *Five Why's*. Kiyo challenged the reader to use the *Five Why's* the next time a problem occurred. That should be easy for us. Problems occurred frequently. Within a few days, I received a phone call from a Plant Manager who loudly told me that our conveyor system had broken down again! This was the fourth time that year that the dumb conveyor had broken down. Each time, it was the same problem. The new micro-controller had a bad circuit board. It was under warranty, but we lost about $30,000 of good glass during the few hours it took to replace it and upload our control programs.

I pulled the computer dude away from his important game and we drove to the plant. I told him we were going to try the *Five Why's* to get to the *root* cause this time. I explained it to him. He was sharp and got it before I finished my first sentence. He said, "That sure beats the *Five Who's*." I knew exactly what he meant.

We arrived at the plant to hear the Plant Manager barking four-letter words. As we entered the plant, the sound of crashing glass drowned him out. Glass is made by pouring sand in one end, heating it to 3,000 degrees and then cooling off the continuous ribbon of glass on rollers for a few thousand feet. You can't turn it off. If the roller system breaks down, beater bars pulverize the good glass into shards, and then we shovel and convey it back to the tank to be re-melted. It was a mess.

We ignored the Plant "Mangler" and asked a few of the good operators to join us by the conveyor. I told them we were going to try the Five Why's. They caught on right away. We would keep asking why until we got an "I don't know" (or a "Homer Simpson"—you know, when he hits himself in the head and says, "Doh!"). Here we go. I hope Kiyo was right.

I said, "The main line conveyor is down."
The team enthusiastically asked, "Why?"
After a moment, I told them that anyone could help answer why.
I said, "Because the rollers aren't rolling."
They asked again, "Why?"
My computer buddy said, "Something must be wrong in this control cabinet."

OK, we're getting somewhere.

The team said "Why?"

My computer buddy opened the door, saw the little orange light, and said, "Dang. The same circuit board failed again."

He pulled it out of the cabinet and laid it on top, then he walked toward the office area to call the vendor again.

I told the team. We need to get to the Root Cause.

I quoted some sage person, "Problems are like weeds. They won't go away until we get the root."

My dandelion-fighting co-workers grunted their agreement.

I queried them again, "Why did the circuit board fail?"

One team member picked up the board and said, "Dang, that board is hot!"

Hmmm.

The team queried again, "Why?"

Someone said, "Maybe the fan's not working."

One worker checked. "No, it's still working."

Then, he exclaimed, "Hey, look here! This little filter is all plugged up!"

Ah ha.

We asked, "Why?"

Someone said, "Because no one changed it."

One more why, and someone else said, "Because no one is assigned to change it, and we don't have any spare filters anyway."

Homer Simpson time! Doh! A quick check showed filters cost about 10 cents each and could be bought by the dozen.

You probably have computer and electrical equipment in your operations. Look closely at it. The equipment is getting smaller. And, it seems to run hotter. Where is the worst place to put an air intake for equipment? That's right, near the dusty floor. Guess where most air intakes are placed? Making matters worse, a glass plant needs to sprinkle white powder between every piece of cut glass or the super-smooth pieces will stick together permanently. Filters near the floor would plug up quickly in this environment.

Without ever having heard of the TPS methodology called TPM, we made our first-ever Operator TPM Form. The workers volunteered to change the filters on the first day of every month, since they were cheap. We kept a small stash of filters by the Operator station. One of the workers drew up a simple *reminder chart* by hand. He misspelled a few of the words, but you get the idea. Once we received the filters and replaced one, he wrote his initials on the Form.

It's fun to do the right thing before you even know what it's called. I wonder if anyone bought and stocked those little filters after I quit working at the glass company.

- Want to do something fun? Go back and count every time you see a word with the prefix re- in this book. Re-anything is basically a "hidden factory" at every organization. Re-work, re-do, re-check, re-draw, re-test, re-paint, etc. Dollars are flowing out the back door of your plant every day these re's stay hidden. This hidden factory is costing you a bundle.
- Does the speed at which people learn to do the Five Why's surprise you? Why or why not?
- Do you hold your engineering teams to their "expected" design targets?
  - Do they even have design targets?
  - Do your engineering teams always follow their designs into the field?
  - Is there a "can't go home until…" criterion used?
  - How might your engineers and designers learn from this exercise?
- Does it surprise you that operators at this factory volunteered to replace the filters? Why or why not? Have you tried involving them in the solution? Are you willing to try it again?
- I am already starting to repeat some sayings like "Problems are like weeds…" Is this okay? In my role (trying to lead a team through root-cause problem solving), did it seem to help?

(Triangle Kogyo) Stand in Circle (9)

"'Don't Just DO Something, Stand There!' Get perspective and reflect rather than just continuing to do the same thing."

Dr. Scott Simmerman (www.squarewheels.com)

My exit from the glass company was a bit unusual. The rest of that story can be found in the People pillar story called "Integrity." I jumped at the chance to work at Triangle Kogyo and learn under a true sensei. Triangle Kogyo made seats for Mazda and arguably had one of the best JIT management systems in our area. The Japanese advisors and Bo Shimono were known as tough task-masters but great teachers.

On my first day at Triangle Kogyo, I arrived early. During the interview process, my General Manager and sensei, Bo Shimono, was in Japan. I was excited to finally meet him. Visions of Mr. Miyagi in *The Karate Kid* danced through my head.

Human Resources Manager, Mike, looked nervous as we approached Shimono's office. We reached the door.

Mike said, "Shimono-san, here is your new Production Engineering Manager."

Then he pushed me through the doorway and ran!

Stunned, I thought, *carpe diem*—seize the day.

I stuck my hand out and said, "Hello Bo. My name is Steve."

The short, stocky man rose and walked out of the room.

As I stood there with my hand still out, I thought, "This is America buddy. We say Hi. We shake hands. We drink coffee!"

Bo poked his head back in the room and said, "Follow me." I stood mutely. It clicked in my head slowly. Follow me. He wanted me to follow him. I had heard only one brief summary of Japanese management styles, in addition to *The New Manufacturing Challenge* book. It was, "Do what you're told!" So, coat still on and briefcase in hand, I raced to catch up with the fast-disappearing Bo Shimono.

Bo stopped in the middle of an aisle as I finally caught up, breathless. He looked up, down and all around. Then, he got on his hands and knees with a piece of chalk. He drew a tight circle around my feet, stood up, loudly pronounced, "Stand in Circle!" and then quickly walked away. I was stunned again. But, I stood in the stupid circle.

8 am.

I need coffee.

At 8:30 am the workers all went on a break.

At 8:45 am they all came back.

9 am. I don't need coffee anymore—I need something else!

10 am. I was fork truck bait, as these smirking drivers swerved dangerously close to me.

At 11 am the workers left for lunch.

1 pm. Hungry.

2 pm.

3 pm.

Still in stupid circle!

Shimono rounded the corner and asked with bravado, "Steve-san, what do you see?"

I dropped my briefcase and raised my hands to strangle the guy who left me out here all day (but did not step out of the circle).

"I can't believe you left me out here all day. I'm hungry. I need to go to the restroom. Fork trucks tried to hit me. I hate this place!"

Bo said, "That's too bad, Steve-san. You come back tomorrow. Also stand in circle."

Argh.

This guy is nuts!

I was desperate. What did Bo want?. As I plopped into my chair at home that night, I picked up Kiyoshi Suzaki's book *The New Manufacturing Challenge* again. Chapter 2 was labeled Workplace Organization. Kiyo seemed to be saying:

You know that fat Training Manual over in the corner—the one with the dust on it? Ignore that! Everyone else does. If you stand there long enough observing, you can learn the real operating system, the way things are really done. **(Steve-san's paraphrase)**

Hey. Maybe Bo wants me to… A cycle that would repeat for the next few months began to emerge in my head. Observation. Check it with sensei. Get redirected with many follow-on questions instead of answers. More observation. More questions. Bo wanted me to observe his operations closely to see if I could detect his operating system. I could do that!

At 7 am the next morning, I was back in the circle. I brought only a pad of paper and a pen. I had just a little bit of coffee (if you know what I mean). Observe. I was near the main seat assembly lines and feeder cells. Yesterday, I was so embarrassed by the 50 or so line workers snickering and pointing at me that I did not even look in their direction. Today, I'm focused on the assembly line.

After a few minutes, I noticed a yellow cord hanging loosely along the entire length of the conveyor line and even the feeder cells on both sides. Every few seconds, a worker would yank on that cord. A yellow light flashed on a scoreboard hanging from the ceiling. The line did not stop. But, a person at a small desk next to each group of seven or eight line workers would jump up and work together with the cord-yanker until they resolved the problem. I wrote all this down and sketched a section of the line. But, every hour or so, someone pushed a different button, or the two workers together could not fix the problem in time. A red light would go off. Wonk wonk wonk! Workers and repairmen immediately appeared yelling, "Fix the line!"

I also knew that we built seats in exactly the same sequence as the Mazda assembly plant up the street. When they slowed down, we slowed down. When they sped up, we sped up. When they stopped, we stopped.

I also noticed that no fork trucks entered the assembly area. Tight aisles were painted different colors through assembly. A guy on a little stand-up electric scooter pulled two trailers of small bins to every workstation on a route—like a milk man—then returned to the storage area. I sketched it on my note pad. He unhooked his trailers full of small, empty plastic bins, and then hooked up to another set of trailers that had full small bins of parts already stacked on them.

Then, the driver guy started his route *exactly* at the top of every hour, just like clockwork. 8, 9, 10, 11, noon. Never early. At nearly every workstation, he picked up empty bins, dropped off full bins and grabbed some laminated cards. It took him about 55 minutes to make his route. He then returned to the storage area and talked to a person there.

Earlier, this other guy pulled empty trailers through the storage area with sloped shelves and picked full, small bins using some sort of shopping list (of cards) and stacked them on the trailers. It looked like a supermarket, so that is how I labeled it on my note pad. (As you probably know, this area is actually called a supermarket.) TPS *namers* know how to keep it simple. Maybe it is supposed to be simple. I called the other guy back there the professional shopper. Don't you wish you had a professional shopper?

As I flipped over my seventh page of notes and sketches, I noticed the time. It was just before 3 pm. Bo Shimono rounded the corner.

He said, "Steve-san, what do you see?"

I blurted, "Bo, Bo, watch that driver guy back there at the supermarket. There he goes, at the top of the hour! He drops off and picks up and... Over there, Bo! Station number five just went down. Guy jumps up. Poke the seat, poke the seat. But, if they don't fix it. Wonk, wonk. Fixers everywhere! And..."

I blurted out seven full pages of observations in one long breath.

I worked for my sensei for quite some time. I never once saw him smile. But, the corner of his mouth turned up and he straightened himself up for a pronouncement. "Very good Steve-san. Last American manager was *two weeks* in circle!"

- Now it's your turn! Walk with me to your operations and let's stand in a circle for a while, even if it is a business process area. What will we see?
  - Will we see a *symphony* of flow?
  - Is everything clearly labeled—so that even Steve-san could detect abnormalities?
  - Will it be clear to us what has been done, and where each object is going?

- If something is wrong, does it stand out like a sore thumb?
- If trouble occurs, is there a *wonk, wonk, wonk* and purposeful scurrying?
- If someone pulls an andon cord and needs help, does anyone *ever* come?
- If no one answers the andon signal right away, does it elevate to a higher-level person?
- Does someone immediately start helping the person or just impatiently deflect input?
- Do you track the number of andon cord pulls? Do you shut down the line and find out why if this number is lower than expected?
- If someone gave you this assignment, would you be able to *see* your operating system?

- What if we stood in the *business* process areas of your operations?
  - Is there a white board next to every person with the projects they are working on?
  - If they are delayed, are the reasons clearly written? If you are the boss, chances are good that you can do something about the delays. Why don't your workers bring up the reasons for delays?
- Should you tell the person in a circle to attempt to observe and document your true operating system, or just let them figure it out? Why or why not?
- Why do you think Triangle Kogyo required raw material deliveries exactly every hour? Is it better to deliver once a day to all stations or every hour? Why?

Note: Stand in circle yourself, then have your key leaders do the same. It also makes a great assignment for a new person. Make sure you take them to lunch after this assignment.

These Always Lie! (10)

"A desk is a dangerous place from which to view the world."

Adapted from *John Le Carre*

"We are what we repeatedly do. Excellence, then, is not an act, but habit."

Aristotle

On my third day, buoyed up by my obvious *success* getting out of the circle, I searched for my desk. I was a bit appalled as I was escorted to a large, open bull pen area. My "desk" was a piece of plywood lying on two file cabinets. My "desk" was pushed against three other similar desks making a larger square. There was only one computer in the middle. It was shared by all four people in this strange setup.

At AE, I had become quite proficient with computers. I started to ask about the setup and computer. Bo quickly interrupted (or erupted), "You do NOT need computer. You go to factory floor. Now, go!" My rapid *circle escape* seemed to have been forgotten. Setting my briefcase and box of personal items on the plywood, I marched out to the floor. I saw a team leader to whom I was introduced earlier. His name was Squirrel (really). He told me everything was running fine.

After an hour, I snuck back to my desk to put stuff away. I tried to introduce myself to the workers around me. Most seemed distracted. But, Mitsumi was nice. She asked me a lot of questions about my background. As I unpacked my box, I noticed that my chrome family picture frames clashed a bit with the plywood. I got a splinter. There was no place to put my volumes of important reference books. Bo Shimono had snuck up behind me.

He said, "Steve-san, how are my assembly lines running?"

I said, "They are running swell, Bo. I just spoke with Squirrel who said…"

Bo erupted, "Steve-san, *these* always lie (pointing at his ears). *These* never lie (pointing at his eyes)."

He said, "You go and see! You go to factory floor!"
My eyes roll.

The first thing I noticed when I stepped into the plant was a crashing sound. I walked toward the sound, quickly found the pack of standers and asked, "What's up?"

Supervisor said, "This press just threw a rod or something."
The tool belts defined the three workers banging on the machine.
I asked, "What is wrong with it?"
Shrugs.
I asked, "How long will it be down?"
Shrugs.

The Japanese advisors next to the press barked, "Quickly is better!" A phrase I would hear quite a bit. After more observation, I got a better feel for who did what. I would ask Shimono-san tonight. Day 3 done; my stuff still in boxes.

- Try this request next time someone says everything is fine or that they need more resources. Say, "*These* always lie (ears). *These* never lie (eyes). You go and see!" What did you learn? OK, for those that actually did this, what did you learn?
- How many of your managers and leaders could survive a "point test"? If you asked them to point to the value-adding persons that they *directly* support, could they do it? If you then asked these workers how many hours per day (if any) this manager or leader *visibly* supports them, what would they say? Can they survive the point test? Can you?
- Where should the offices of managers who directly support production be located?
- In the eyes of your supervisors and leaders, how can you increase the priority of the gemba—the place where work occurs?

Time Study—Part 2 (18)

"Tell me and I will forget, show me and I may remember, involve me and I'll understand."

Chinese Proverb

Remember when I vowed that I would never do a time study ever again while at AE? Well, here comes the rest of the story. A few weeks after the reverse engineering event (see my People pillar story called "You Bad Guy—Reverse Engineering"), Bo called me into his office, threw a stopwatch at me, and said, "Do Time Study on seat track run-in station!" Oh no! Not a time study. In case I didn't mention this, Triangle Kogyo was a unionized company. It was the same union as AE. Visions of my last time study a decade earlier haunted me.

I drag my feet when I do not want to do something. Just like my teenagers. I was really dreading another one of these *games*. I found a blank sheet of paper and finally rustled up a clipboard. It was no Picasso painter's palette. It looked like it had been run over by a fork truck, with a corner missing! I looked at the stopwatch. It was in decimal minutes (DM). What the heck is a decimal minute?! I dragged my lousy time study materials toward the seat track cell.

About halfway to the cell, my friend and team leader Squirrel walked up beside me and grabbed my stopwatch.

He said, "Where are you going with that stopwatch?"
I said, "Bo asked me to do a time study."
Squirrel said, "*You* can't have a stopwatch on the floor."
I said, "Why?"
He said, "Because you're management. Union workers are the only ones that can use a stopwatch out here."
I was perplexed.
I said, "Bo told me to come to the seat track cell..."
Squirrel interrupted and said, "Oh, I see now."

Squirrel continued, "We asked for help on the seat track cell. You see, a second shift worker on the run-in station thinks she has a better work method than the person on first shift. We just wanted some help judging whether her safety and quality are better. You do know something about ergonomics, right?"

I asked, "Could you please repeat that again slowly?"

Squirrel rolled his eyes and repeated slowly, "We didn't ask for a time study. Union members do all of our own time observations and line balancing here. My second shift worker *found a better way* to build than the current documented standard. We just wanted someone to ensure that it is as safe and high quality as before. You're an Industrial Engineer. You can do that, right?"

Remarkable.

I just nodded. My head was spinning. I wondered if Bo knew this team did *not* want a time study. Why did he ask me to do one?

As we neared the seat track cell, Squirrel said, "No need for the stopwatch. We know it's faster just by using our wrist watches. Just watch her for a few cycles, we'll videotape her, and then come back tomorrow to watch first shift doing it the current standard way." I realized two important things that day:

1. Unions are not the problem.
2. When given the tools and authority, people will come up with simple ways to do things better. It's called the *Law of Least Effort*. I call it LOLE (sounds like lowly).

People will find a faster way. It's in all of us. Most companies don't even try to tap it. What percentage of your creative brainpower, in the form of ideas, is your company asking you for and implementing daily?

- Who owned, used and controlled the tools of improvement (time study) at AE?
- Who owned, used and controlled the tools of improvement at Triangle Kogyo?

(continued on next page)

- What difference did that seem to make for the workers?
- How would an improved work method or sequence benefit the worker? How about the company?
- What concerns might a worker have with reducing the overall operator cycle time?
- How secure in their jobs did the Triangle team members seem?

Dead Flies (24)

"This may be a perfect opportunity to use common sense!"

Robert Pike

Bo Shimono used many funny idioms and analogies. One of our teams did a 5S (workplace organization) exercise. The first S is Sort/Scrap (*Seiri* in Japanese). You are supposed to get rid of all junk and waste first. Our team members skipped that step and went to Step 2, "Straighten" (*Seiton* in Japanese). They labeled everything—even the junk and waste.

Bo reviewed our progress. He frowned, grunted and then said, "If you see dead flies on window sill, clean them up! Do not label them!" This one does not need an explanation. (See Figure 2.1.) We carted out a full dumpster of unnecessary items that week.

- Think of all of the labels in your facility. Do any of them amount to labeling waste? Which ones? What could you do about it?
- When people label waste, which step are they skipping? Why?
- Are there any other seemingly wasteful tasks or actions done in the name of Lean or TPS? Why do you think they remain?

Figure 2.1   Dead Flies.

(JCI/Toyota) You Will Fail (26)

"Progress is impossible without the ability to admit mistakes."

Masaaki Imai

JCI was starting up seat assembly and parts plants quickly. They had an assembly plant and a parts plant serving Toyota that performed superbly. But, they also had dozens of others that made junk fast. The Vice President of Manufacturing was a large, boisterous man. I liked him. He asked me to figure out what made our Toyota plants great, document this in the form of an Operating System, and then spread it to the other seat plants. He said I had one year to do this. Desperate to continue my learning from Toyota, I said, "Sure boss, I'll do it."

I booked a one-day trip to Toyota. Oh boy. I met the man from Toyota who would be my sensei, Hiroyuki Nohba. I was also to meet with my advisor within JCI, Phil Beckwith. Some expensive universities wrote case studies about how Phil was pulled, kicking and screaming, into the Toyota way. I liked him right away. Nohba-san. Well, that was a different story.

Looking at my watch, I asked THE question. "Mr. Nohba, will you please tell everything you can about the Toyota Production System?" Hiro frowned. Yikes, when he did that, he looked like...

Figure 2.2   Toyota House Model, adapted by Steve Hoeft from Jeffrey Liker, 2002.

Concentrate.

Get your pen ready.

Hiro drew the familiar foundation, three pillars and roof of the TPS (see Figure 2.2 below).

Hiro paused, searching my face.

Then, he briefly explained the foundation of operational stability.

I nodded excitedly and said, "Got it!"

He frowned.

He then explained the first pillar, JIT.

I interrupted quite a bit with my *deep* knowledge of cells but told him, "I still need to learn more about kanban."

He frowned and looked for a moment like he was going to leave. Phil's pleading eyes seem to calm him.

Hiro explained the second pillar that he labeled *People*.

I nodded and said, "Got it!"

He frowned.

My eyes darted back to the JIT pillar.

After a lengthy silence, Hiro briefly explained the third pillar, Jidoka (using some literary privilege, we will call this *Built-in Quality* [BIQ]).

He said, "Pillars must be built with equal intensity and at the same time."

After a pause, he said, "You cannot put the roof on the house until the walls are built. And, you cannot build walls until you have a strong foundation."

He declared, "This is TPS."

Marveling at the simple drawing, I blurted, "I got it. Foundation, JIT, People and Quality."

He simply said, "You do not."

I was hurt. Hiroyuki did not know about my intense study of all things Toyota and time with a sensei like Bo.

I countered, "I think I get it. Foundation, JIT, People and Quality. It looks easy!"

He said, "It is not. In fact, you will *fail!*"

I said, "Fail! I can't fail. I mean, we must succeed, or the whole company is in the toilet."

Hiro said, "You will fail."

I was not confident enough to press Hiro further. So, I said goodbye, set up a longer follow-up meeting in a few weeks, and flew back to Detroit. Failure. How could I fail?

For most industry leaders, the desire for TPS-like (or Lean-like) results is usually stronger than the desire to build the disciplined systems required. This is not a good thing. In other words, most leaders do not have the guts to put the operating system principles in place and sustain them. TPS is a "contact sport." You must trust your people, develop your people, and then build systems around them for continuous improvement. Leaders tend to have the attention spans of a gnat.

I returned to JCI headquarters. By the end of that same week, I was pulled into a metal seat track plant that was struggling to meet production goals. With the pre-work done by a great young engineer, we rapidly moved the small welders, assembly equipment and fixtures into a tight, five-station, one-piece flow cell. It took less than a day to do it. The engineer and I were near tears (of joy) as we left that night. Our cell looked like lean. It even smelled like lean.

When we entered the plant the next morning, we were quickly dragged into the Plant Manager's office. A Plant Manager's qualification must be the ability to yell, spit and swear at the same time. Even with our constant interventions, our beautiful one-piece flow cell was broken down more than half of the day! All of the workers and support staff were shut down. Our cell looked like lean, it smelled like lean, but it ran like crap! The Plant Manager screamed while I called Nohba.

I called Nohba, "Nohba-san, help! We put together a perfect one-piece flow cell at our seat track plant, but it was broken down more than half the day! Help!"

I could hear him smiling through the phone.

He said, "Steve-san, you have failed."

Great. That is the support I needed.

I said, "I know, but how do we fix it?"

He simply said, "Change back."

I said, "Huh. Why?"

He said, "Steve-san, you have forgotten the **Foundation**."

An image of the Toyota House flashed in my mind, with his prophecy that I would fail.

I said, "I will put the machines back where they were. Please tell me where I have failed."

Hiro asked, "What is the percentage of time Machine 1 is up and available to you?"

I said, "It's a typical welder. I think it's 88% to 90%."

He said, "No think. Get numbers. Call me back."

After a mad scramble and orders to smirking skilled tradesmen to move the equipment back, I rounded up the best downtime data I could find.

I called Hiro.

He asked again, "What is the percentage of time Machine 1 is up and available to you?"

I said, "This welder is down 14% of the time, mainly for weld tip and wire roll changes, so it's 86%."

He said, "Machine 2?"

Welder 2 was up 87%, Machine 3 was up 90%, Machine 4 was up 91% and Station 5 was up 87%.

Hiro asked me to multiply the *uptime* of these machines which I had moved in series (versus in large batch departments or islands) to get the net.

Darn.

The net cell uptime number was almost exactly what we experienced.

He said, "Steve-san, you have failed because you have forgotten the foundation. You skipped forward to one-piece flow cells because it looks sexy. **Preventive Maintenance** is in the foundation. You must address your downtime issues before you can build one-piece flow cells!" Wow. He was right. The downtime numbers did not seem to hurt us when we produced in batches in over-capacitized departments. But, in a lean system, this amount of downtime was fatal. I wouldn't make that mistake again!

Well, I made that same mistake at least four more times before I learned the importance of the foundation. The second time, it was the Foundation principle of First Time Quality (*Robust Processes* on the Toyota House). The next time it was early *Supplier Involvement* on a new launch.

Dang.

Hiro had guessed these failures.

It's a good thing Toyota is patient.

Though, I did hear Hiro repeat, "Quickly is better."

- Has your company made this mistake—attempting to install cells, kanban, TPM systems, or BIQ systems—before buttoning up the foundation? If so, what did you learn? Was it worth it? How do you prevent that type of mistake next time?

  (continued on next page)

- The Toyota House Foundation is made up of *continuous* improvement principles. Since you must work on these continuously, when are you ready to move on to the pillars?
  - One hint from Hiro: Stability. It's a principle.
- Determine the total percentage of planned production time that machines in sequence are "up" and available to you. What is your expected system uptime? Are you collecting all of the downtime with reasons? Why or why not? How do you collect a single "slow operator cycle" or two?
- Was it right to "change back" the cell until JCI addressed the foundation? Why or why not?
- What would you say to a consultant or leader who says, "You must implement kanban everywhere, now!"? Assume you had some foundation principles to work on. Would your leaders be willing to wait?
- When do you move on to the bigger tools of JIT and BIQ?

Timeless, Unchanging Principles (27)

"Never mistake motion for action."

Ernest Hemingway

Hiroyuki Nohba from Toyota also convinced me of another important facet of the Toyota House. He called the words in the foundation and at the top of each pillar "Timeless, Unchanging Principles." Hiro said, "We *always* implement the principles. They *always* work. We will *not* compromise on these." After many bad attempts and then many successful TPS implementations, I started to fully appreciate the principles. When I compromised principles, we tended to get stuck in the mud, or at least slowed the pace of change. When I stuck to the principles, we achieved flow with better quality and even improved morale.

The best example of timeless, unchanging principles is kanban. Later in my career, I served as a lean coach for a muffler manufacturer. The facility had been told by a high-level Vice President to install kanban between every assembly cell and their raw or Work in Process (WIP) cells. So, they did. They printed lots and lots of kanban cards. So many in fact, that they were used like disposable labels.

The principle in the Toyota House is *flow*, and when you can't flow, *pull*. Going directly to a *tool* like paper kanban actually prevented this facility from transforming directly to one-piece flow. And, the best way to pull was not paper kanban in most of these loops. Simple, visible kanban squares (outlines on the floor) or a two-bin system (very small number of well-labeled containers or carts) would have been better than kanban cards.

By the way, this plant stopped using kanban after a few months as a true *permission to produce* requirement. They kept losing the kanban cards, even though hundreds of cards kept floating around in the system—just in case the Vice President wandered through.

- What are the timeless, unchanging principles from the Toyota House?
- If a high-level leader says to implement kanban (not try one-piece flow first), what would you do?
- Give me one example from your operations that you could link together with kanban. Is it possible to achieve one-piece flow between them?
- If a factory does not have time to start building foundation principles, is it a compromise to skip foundation problems and move on to the JIT tools? Is it a compromise to avoid attempts at flow and pull? Is it a compromise to avoid attempts to "never pass on a bad part"?
- Do you use pull signal or kanban as a true permission to produce requirement?

One-Tool Only Consultants (41)

"If all efficiency experts were laid end to end—I'd be in favor of it."

Al Diamond

Hopefully they haven't gotten you already. Some large lean consulting companies like to sell 5S workshops first. Now, there is nothing wrong with 5S. I use it every time we implement the TPS principles. I just do not implement 5S *alone* any more.

Some say, "There is no way you can sustain ANY of the other lean tools if you can't sustain 5S first." Note: there is some truth to this statement. So, they sell you a bunch of five-day-long 5S workshops at $20,000 each.

They gather your people together from every department 10 at a time. They swoop into the next area, then the next area. By the end of the year, the first few areas have "reverted" back to their old ways. Then, the unscrupulous consultants try to sell you ANOTHER round of 5S workshops! Sadly, many buy this load of hooey. In many cases, management votes to stop the "Lean" effort because it does not have a high enough payback. Except for the consultants, that is.

And, who knows where the months of inventory went by the end of the fourth or fifth day of their 5S workshops—it somehow left the area they were focusing on that week. I think they just moved some inventory to another location, only to move it right back during the next 5S workshop. It was like the carnival game of *whack-a-mole*. You hit the little plastic mole with a mallet, but he just pops up somewhere else. It's a fruitless game. You can't win!

- How about you? Has some consulting company tried to sell you only 5S workshops?
- If so, did the consultants offer to teach your staff how to conduct the next similar workshop?
- Have you tried to implement 5S without any other tool or principle?
- Have you tried to do 5S that reinforces a key change? I bet it worked better in this instance.

3

JUST-IN-TIME PILLAR PRINCIPLES AND STORIES

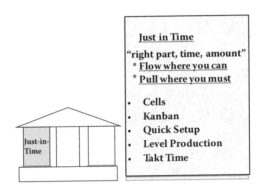

Just in Time
"right part, time, amount"
* Flow where you can
* Pull where you must

- Cells
- Kanban
- Quick Setup
- Level Production
- Takt Time

Just-in-Time

In housing construction, you build a strong foundation first. Then, you put up the walls concurrently. You can stand up one wall, but you will need to spend time propping it up until all of the other walls are attached and square. This is true with all three pillars of the Toyota House both in sequence and dependence on each other.

We will start with the Just-in-Time (JIT) pillar. All pillars are important, but the JIT pillar includes some of the most powerful principles in the Toyota Production System (TPS) arsenal. The timeless, unchanging principles in this pillar are JIT—the right part at the right time in the right amount—flow and pull. There are also several tools listed in this particular version of the house—cells, kanban, quick setup (aka Single Minute Exchange of Dies [SMED]) and takt time. The tools are often tailored and adapted to a particular area and may be different, depending on the process.

JIT production is making what the customer needs, when it is needed, in the quantity needed. This must be done using the minimum resources of manpower, material and machinery. Its main purpose is to combat two main wastes—overproduction and inventory.

In a traditional production system, inventory exists in case problems arise. In a JIT system, the objective is to *eliminate* inventory. JIT systems require small lot sizes, low setup times, close physical proximity of processes and containers with a fixed number of parts. Thus, in a JIT system, there is only the minimum amount of inventory to keep the system running.

What does a JIT system do? It achieves streamlined production by reducing inventory, thus reducing waiting times. It allows for a more balanced, smooth production of goods and services. A JIT system also attacks waste. Any tasks or steps that do not add value to the product or service from the customer's perspective are highlighted in a JIT system. It exposes problems and bottlenecks caused by variability. Any deviation from the optimum is exposed and corrected. Some people believe JIT systems are brittle because you are forced to stop and correct the defect or problem. Toyota believes that it costs your operations much more by NOT stopping!

Some of the key principles in the JIT pillar are:

- Right part at the right time in the right amount.
- Continuous flow—Like a stream, processes add value and then keep the product moving without stopping.
- Pull—When one-piece flow is not possible, pull systems link the customer to the supplier process; pull systems merely replenish what was just used in the smallest batch sizes possible.

JIT or the "Rights"

Making things JIT is a principle. It always works. It means to make the right part at the right time in the right amount. Never early. Never late. Kiichiro Toyoda, the founder of Toyota's automobile business, developed the concept of JIT in the 1930s. He decreed that operations would not make anything before it was needed. He declared that no one would make a single part unless they had *permission* to produce it.

These were not decrees forged out of logic or even lots of thought. Kiichiro and the fledgling Toyota automotive group adopted JIT because it was the ONLY way they could produce vehicles. The island nation of Japan was not blessed with all of the natural resources needed to make cars. They had only one option. They needed to take orders (and deposits) from customers, order components and quality-tested modules from suppliers and then assemble cars as quickly as possible. They delivered quality vehicles to the customer and received full payment in many cases in time to pay their suppliers with money just collected from the customer. If Toyota could order components and build cars before they needed to pay their suppliers, they would not need to borrow money. They could make money using their customers' money.

It was out of sheer necessity that the greatest addition (JIT) to the greatest production system was formed. There is no other system that can achieve flow and generate cash like a JIT system! Under Ohno's leadership, JIT developed into a unique system of material and information controls that was much simpler than the typical U.S. managed and scheduled operations.

JIT is not a zero inventory system. A zero inventory system would require equal process cycle times, highly stable processes, close physical proximity of processes and little or no changeover time. We may have some operations today for which we need to make components a few days early. But, as we apply TPS principles, the amount made early should decrease dramatically. It is possible to make every component in smaller and smaller quantities. And, the amount made late should reduce to zero. The beauty of the JIT system with low strategic buffers of stock is that the system desires to stay full through rapid replenishment systems versus the large stock "push" production systems that are almost always running out of some components.

I have heard the question before, "Why do I need to wait until I make my component? Eventually, it will be needed, right?" The answer is, "You must wait." Making something before it is needed is the waste of overproduction. It is always the wrong thing to do.

Making a small amount of safety stock does not seem bad. But, on a bottleneck machine, you just made components for some assembly that cannot be shipped today *instead of* making something that could be shipped. We wasted our capacity on something else, just because it was easier for us to make more subcomponents today.

But worse, when we allow workers to overproduce, they are not doing the most important part of a pull system—stopping to help others! The parts they overproduce will not allow us to meet customer demands today. Instead of making more parts (and burying their internal customer with parts), they need to stop what they are doing, walk down to the internal customer and help! The right thing to do is stop, and help.

Making products in advance can also cover up many other problems and wastes. When you have too much inventory, parts wait, people wait, workers shuffle inventory around, defective parts remain covered up, and staff gets redirected to find parts and count stuff. I have seen stock handlers search for specific parts for hours. The computer inventory system says we have the stock, but we can't find it. Watch your fork trucks for a few hours. How often are they driving around with empty forks looking for something?

Continuous Flow

A continuously flowing stream does not stop or pool up. It keeps moving. The same is true in production and above shop floor systems. The flow principle seeks to produce and move one item at a time (or a small batch of items) through a series of processing steps as continuously as possible, with each step making just what is requested by the next step. No stops. It is also referred to as one-piece flow, single-piece flow, and make-one-move-one. Flow is the goal. If you look down into your toolkit of TPS principles and tools, flow is the biggest tool. It would be like a large reciprocating saw that sticks out both ends of the toolbox. It helps you keep the product moving versus sitting.

A good mantra describing this principle is "Flow where you can and pull where you must (and never push)!" Flow is the goal. Many times, I have seen the impact of leaders edicting kanban cards into a production system. Instead of changing the layout to achieve flow, workers just print (lots of) kanban cards. Kanban is supposed to control and limit production. Allowing workers to print extra cards defeats the purpose. And, these factories rarely attempt one-piece flow, once kanban is in place, because they feel they are already *lean*. The point is, keep trying to achieve one-piece flow until you run out of brainpower.

Small lot production is sometimes used synonymously with flow. Small lot production is the process of economically producing a variety of things in small quantities rather than producing things in large batches. Please read the story "Kanban Is Evil" for more details on one-piece flow.

Pull

Pull is a unique principle and is linked to flow. Like other principles in the house, they are dependent on each other. If you can't achieve flow, you must install some sort of pull system. The other alternative is to push products to the next process without checking to see if they are needed. The push or "scheduled" production system will fill all available space with overproduced parts and inventory. Push systems usually exhibit quality problems and other forms of waste. Without pull systems, you will overproduce.

In a pull system, the work performed at each stage of the process is dictated solely by demand for materials from the immediate next stage. To differentiate, in a *push* system, material is pushed into downstream workstations regardless of whether resources are available or material is needed. In a *pull* system, material is pulled to a workstation just as it is needed.

A good example of pull systems is a simple painted outline where a cart or container of work in process (WIP) parts can be stored in front of the next process. If the outlined square (aka kanban square) is empty, the supplying process will fill it with the proper parts until it is full. If it is full, then the supplying process will not produce more. The worker waits for a signal to produce something else, or he or she goes to the customer process and tries to help. Even in non-manufacturing processes, pull systems simply prevent you from over- (or under-) producing. It forces the worker at the supplying process to ask, "Is my internal customer ready for my work right now?" If the answer is no, the worker will not produce any more until their pull system signals for more.

Early supplier involvement was in the foundation. But, a case could be made that it needs to be in the JIT pillar as well. This case could be made with several other principles as well. It is absolutely crucial to ensure the immediate availability of all parts that go into the final assemblies in order to keep flow going.

Takt Time

A critical element of JIT is takt time. Takt time is the pace of production needed to meet customer demand. The calculation for takt time is the net available time for production in a period (like a shift or day) divided by the customer demand in that same period. If a factory had 7.5 hours available each shift and customers demanded 250 parts of a certain model each shift, then the takt time is 1.8 minutes per piece (7.5 hrs × 60 min/hr/250 pieces). This means that your production line or cell must make a part every 1.8 minutes in order to meet demand. Used with the JIT principle, this means that we do not want to make parts much faster than 1.8 minutes each, and that each worker should be working just under 1.8 minutes on each part on the average, if the workload is balanced.

Takt time can be used to design and balance tasks for individual workers. It can also be used to design cells and lines.

Cells

Cells are a specific form of one-piece flow that reinforces JIT. Cells are a tailorable tool, but worthy of some discussion here. When products are created one by one in a cell, lead time is made as short as possible. The customer is pleased, and the rate at which the system generates cash is maximized. A cell does not necessarily require a U-shape layout and counterclockwise flow. But, that can be ideal in some medium- to high-volume production systems. Business process and knowledge-worker (like new product design) cells can also be designed to accomplish the same goals. When workers get within sight (visual management) of each other, productivity, quality and communication all improve. When a team creates a complete product for a customer without leaving the area, it creates great job satisfaction with minimal lead times.

Kanban

Kanban is roughly translated "signal card" in Japanese. It has its root in the simple replenishment card that a stock boy uses to restock the grocery shelves. Like cells, it is a tailorable tool, but worthy of some discussion because of its usefulness in achieving JIT. Kanban is a specific form of the pull principle. (Note: Kanban is pronounced 'kahn-bahn'.) It authorizes production from downstream operations. No one can start production without this permission to produce. It basically pulls material through the plant.

A kanban may be a card, a flag, or even a verbal signal. It is often used with fixed-size containers. It is possible to add or remove kanban cards or signals when the total demand changes. Some people mistakenly feel that kanban is *the* solution for material flow. They feel that more kanban cards make a system *leaner.* Actually more kanban cards make a system *fatter!* Each kanban represents inventory, which is waste. They are merely part of a system that pulls materials to their point of use, and the goal is to minimize the inventory required.

Quick Changeover or SMED

Dr. Shigeo Shingo, a professor of industrial engineering and later a consultant to Toyota, developed Quick Changeover or SMED principles over a 20-year period. The goal of completing all changeovers within one minute, or at least a single-digit number of minutes (ten or fewer for the most complicated machines), is a great target. And, it is possible! SMED has proven its effectiveness in many companies by reducing changeover times (non-value-added or lost time on critical capacity machines) from hours to fewer than 10 minutes.

Employees at Toyota are responsible for their own setups. This helps to reduce waste because the entire changeover process is non-value-added to the customer. Changeovers take up valuable equipment and labor time. Once the time is wasted, you can never get it back.

Like cells, SMED is a tailorable tool, worthy of some discussion because of its usefulness in achieving JIT. SMED seeks to reduce the lost time between value-added processing steps. It seeks to reduce the lost time due to changeovers. It states that you can always reduce setup cost further by reducing setup time. Setup time reduction is a prerequisite to lot size reduction.

SMED is a simple five-step method in which the first step is to list all of your current setup tasks and mark them either "I" or "E" for "internal" or "external" processing step. An internal setup consists of setup activities that *must* be performed while the machine is stopped. An external setup consists of setup activities that can be carried out while the machine is still operating. The second step is to convert as many currently internal tasks to external tasks (either before or after the machine is idle). Then, the third step is to reduce internal tasks further, the fourth step is to reduce the external tasks, and the fifth step is to try to reduce the changeover altogether.

My Most Interesting JIT Stories

The stories that follow will give you a better feel for the **JIT pillar** tools. Please read these and reflect on the questions that follow each one.

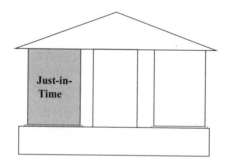

Where Is... Factory? (33)

"Managers will try anything easy that doesn't work before they will try anything hard that does work."

Jim Womack

I was invited to attend a workshop in Indiana led by a consulting company that I will call Takasui. They charge a lot of money to conduct a week-long kaizen event, which is worth every penny. By the way, they will get back into their limousine, fly back to Japan, and charge you the full amount if your site leader leaves them for even a minute. The workshop leader was an ex-Toyota Plant Manager named Koji.

As the nervous Plant Manager of the automotive supply company paced in front of the projector, Koji got out of the limousine. The Plant Manager's administrator tried to gracefully usher Koji into the conference room where donuts and the factory's "success story" slides awaited him. But, Koji was not there to eat donuts and listen to hype. He walked past the conference room, briefcase in hand, and headed toward the shop floor.

The Plant Manager yelled, "Don't just sit there, follow him!" So, the mother duck and baby ducklings formed a line as we followed the skinny, older Japanese gentleman toward the factory. The Plant Manager wormed his way to the front of the line and began spouting the good things his plant was doing while trailing the fast-moving Koji. He spoke of United Way donations, Boy Scouts meetings in the front lobby, quality flags earned on his watch, and so on. Koji opened the main double doors to the plant and stopped suddenly. The whole line of ducklings ran into each other. The Plant Manager exclaimed, "Welcome to my beautiful factory!"

Koji looked long and hard at the factory from where he stood. Then, he asked in a thick, Japanese accent, "Where is… factory?"

The Plant Manager thought to himself, "Tens of thousands of dollars for this! He barely speaks English and is staring at my factory asking me where it is!" What do we do in the United States when someone does not understand us? Yes, of course, we speak *louder* and *more slowly*. The Plant Manager, let's call him Joe, replied, "*This is my **factory** Koji, this is where we work*. It is x thousand square feet and x hundred people work here. This is my (very loud) **factory!**"

Koji asked as if he didn't even hear him, "Where is… factory?"

Joe rolled his eyes.

How patient would your boss be at this point?

He repeated with even greater emphasis, "*This is my **f-a-c-t-o-r-y***. Koji, do you see that corner of the factory to your left?"

Koji responded with a simple "hai."

[**Note:** This may only mean your guest heard you; please do not mistake this word for agreement with what you are saying.]

Joe continued, "All the way to that back corner are raw materials, raw materials. Do you see that back right corner?"

"Hai."

"From left to right, behind those racks, are all of our molding machines. That is all Work in Process (WIP), Work in Process. Do you see that closest corner to your right?"

"Hai."

"From the back corner to the truck dock, behind those racks, is our final assembly line and finished goods. There is a truck. Vroom, vroom. Bye, bye."

Joe again spouted off the total square footage and employees.

Joe said one more time, with great emphasis, "*This is my f-a-c-t-o-r-y!*"

Koji stated his question exactly the same way again, "Where is… factory!?"

Joe looked like he was going to cry.

He whined, "Koji, I keep trying to tell you, over and over. This is my factory. You don't believe me!"

Koji said very clearly, "I see no factory! I see big warehouse. And, you are not very good!"

[Observe] He was right. From where we stood, you could not *see* a single value-added machine or worker. As far as our eyes could see, there was only rack after rack of inventory, which is waste. Fork trucks were roaming around with nothing on their forks, and some drivers even appeared to be sleeping. From where we stood, it appeared that we were all in a very poorly run warehouse. Koji was telling the truth.

Back at the hotel that evening, after dinner and karaoke, I asked Koji if he would teach me to be a TPS Facilitator. Koji said, "Steve-san, you funny guy. I teach you a very technical TPS tool. I call this the *squint test*." Koji put the plant layout on the floor, highlighted the value-adding machines and assembly lines in yellow. Then, he stood back and said, "Now, we do the *squint test*." Perplexed, I asked,

"What do I do?" He gave me an impatient look, and demanded, "You squint at the drawing. Squint with your eyes."

After a few moments of staring at the highlighted layout, I had to ask, "What am I looking for?"

Koji looked at me as if I were a moron, and then stated simply, "Yellow."

[Pause]

Oh, I see!

Koji said, "Until this factory is more than 75% *yellow*, we have nothing but lousy warehouse."

Koji wanted me to size up the layout with my eyes. Just like all TPS or lean tools, Koji was trying to teach me to identify waste. Just like Value Stream Maps, or pie charts of value-added work in a day, Koji was teaching me a key assessment principle of TPS. Even floor space can be measured in terms of *percent value added*! Even floor space must become more value-added every day. What a lesson!

Note: If you went to this Indiana factory today, you would see that they increased sales by nearly 10 times without adding one more square foot of space or even one more worker!

- Application and reflection: Highlight your plant layout in the way described above—only the value-added equipment and areas. How *yellow* is it?
- What is your percent value added of floor space? If you set a goal to double it, where would you start?
- If you had the floor space, people and some equipment freed up, how hard would it be to generate some additional sales? What needs to happen first? Do you have sales growth in your plans?
- What is your percent value added for all key process tasks for a single product (dock-to-dock)? Do you use Value Stream Maps to help uncover this?
- What is your current revenue or profit per square foot?

Must Cut Inventory in Half, or a Bias for Action (34)

"I have been impressed with the urgency of doing. Knowing is not enough; we must apply. Being willing is not enough; we must do."

Leonardo da Vinci

When last we left Koji, he was standing at the door to the factory, briefcase in hand, saying, "Where is factory?" He walked quickly to the back of the factory, as if he did not believe that *any* value-adding machines or processes existed here. After seeing the molding machines, he relaxed a bit and then looked at the rows of four-high pallet racks filled with tons of molded WIP inventory. Joe said one of the dumbest things I ever heard. He said, "I see you are admiring my well-labeled WIP racks. We 5S'd all of our inventory!" [**Note:** See the story "Dead Flies" in the Foundation section.]

The manager had dedicated space for each WIP part number in these racks and labeled them very well. When you dedicate space, you need enough space for the *maximum* number of parts that would ever be stored in a one-year period. All inventory is waste and should be reduced, not labeled.

Joe proudly narrated as he walked down the aisle, "Work in process inventory from the molding machines is stored in all of these racks. We keep all WIP for Part Number 123 here; all Part Number 125 is here..."

Koji was not following him.

He was staring up at the first rack.

He asked, "All Part Number 123 is here?"

Joe said, "Yes. We keep all WIP for Part Number 123 here."

Koji said loudly, "Must cut inventory in half."

The Plant Manager might have been thinking to himself, "That's what we are paying you to teach us."

Joe said slowly, "Yes. That's a good idea."

Koji seemed to be waiting for a different response.

He repeated, "Must cut inventory in half!"

Joe looked perplexed and said, "I heard you the first time."

The only way to describe what happened next is the word *sumo*. Skinny Koji slapped one foot down, then the other, saying, "Must cut inventory in half!" The Plant Manager, already embarrassed earlier, was not about to be embarrassed again by this skinny man. He slapped his feet down one after the other and screamed, "I said that I heard you the first time." Koji smiled. It was clear he was enjoying this. Then, Koji screamed as he stamped his feet and repeated several times, "Must cut inventory in half!" The Plant Manager said as he pulled a notecard out of his pocket, "OK, OK, Koji. Calm down. I'll write it down."

(Faster) Koji threw his briefcase on the floor, pulled out a saws-all (metal-cutting reciprocating saw), plugged it into the column next to the inventory rack, reached up as high as he could, and started to cut the support post of the storage rack filled with tons of loaded pallets. Joe turned in circles as if he was going to grab Koji, then call for help, then grab Koji. It was a surreal scene. I backed up thinking, "That rack will fall; that guy's going to die; this is called kaizen—this is funny!"

Joe tried to pull Koji away, but Koji just growled back and kept cutting. Eventually, Joe eased Koji away from the full racks and then called a fork truck driver to empty the top two rows of pallets from the inventory rack. Koji stood aside, still revving the motor of the saws-all. When the last pallet was moved aside, Koji smiled and then continued to cut through all four posts of the rack. We all watched as the fork truck driver carefully moved the top of a well-labeled Part 123 inventory rack outside of the plant. When he was finished, Koji pronounced, "Inventory in half! We may now begin the kaizen event!"

Later that night, after the squint test, I said to Koji, "You could have died out there."

He said, "No, Steve-san, I trusted that he would stop me."

I joked, "You embarrassed him twice in just 10 minutes. I think he would rather see you dead!"

Koji said, "Steve-san, I do this *every time* I enter a factory! It is my *signature*!"

[**Note:** If you see a Takasui consultant, check his briefcase!]

Koji asked me again, "Do you *really* want to be a TPS Facilitator?" I told him, "Yes. Yes, I do."

Then, he told me, "Then you must be crazy man! Americans get 'glue in seat' during kaizen activities. Always sit and eat the donut; eat the donut. You must *lead* them out of their seats to the shop floor where the work really occurs."

He continued, "You must have a **Bias for Action**, or nothing will get done."

He was right.

Wait until you hear what Koji did at his next workshop!

Note: A quote from my sensei, Bo Shimono, "Quickly is better."

- Application and reflection: Can you cut some of your inventory storage racks or areas in half? Why or why not? Why not do it today?
- If you physically constrain the only locations where you store inventory, how would your production systems react? Would you need to audit these areas frequently to prevent overproduction?
- Do you have plans to reduce all inventory levels? How are they progressing?
- Is it true that participants in a kaizen event prefer to stay in a conference room, sometimes eating donuts? Why? What can you do to lead them to the place where work occurs?
- Remember, safety first.

Door Here! (35)

"The impossible is often the untried."

Jim Goodwin

About four weeks after the infamous rack-cutting incident, I received a call from the TPS coordinator at the Indiana factory.

She said, "He's back!"

I said, "Who?"

She said, "Koji."

I said, "Did you check his briefcase?"

She said, "Yes, nothing but papers this time."

I said, "What happened?"

She said, "Do you remember how the molded rubber parts are moved from the press to the deflashing cells all the way around the wall to the old extension?"

I said, "Yes."

She said, "Koji was doing a 'Jeffy Walk' from Press #1, around the wall, back to Deflash Cell #1..."

[**Note:** Every TPS tool attempts to make *waste stick out*, so that team members can reduce it. One TPS tool is to graph the walking steps that a person or material takes with dashed lines on a layout of the plant. Some TPS Facilitators call this a "spaghetti diagram" since all the overlapping lines look like spaghetti on a plate.]

Others, like me, call it a "Jeffy Walk" after a repeated theme in the cartoon "Family Circus" (copyright King Syndicate). Wherever Jeffy or his big brother Billy walk, they leave a trail of dashed lines across the cartoon strip. It can be funny. I challenge you to draw worker and material movements for every key process in your factory. Jeffy is in your factory! We must find him and get rid of him, because excess part travel and worker motion represent great wastes in a lean production system!

Koji was doing a "Jeffy Walk" between the rubber molding and deflashing machines. He walked from the rubber molding press all the way around the wall to about the same spot behind the wall. He stopped and then retraced his steps, counting in Japanese this time as he returned, "Ichi, ni, san, shi, go, roku, shichi..." He stopped at the opening along the far wall, looked back to the second set of machines, and then started counting again along the wall all the way back to the first machine. His count ended on just about the same number. He stared at the wall. The operations were only a few meters apart but separated by that wall.

Koji said, "Door here."

The Plant Manager said, "You're right Koji. We should put a door here."

Koji paused then repeated, "Door here!"

The Plant Manager looked perplexed again, and said, "That is a good idea Koji. We plan to do just that."

Sumo! Koji repeated for a third time stomping his feet, "Door here!" The Plant Manager made the mistake again of repeating that he heard him, then said as he pulled out a notecard in his pocket again, "OK, OK. I'll write that idea down."

(Faster) Koji jumped on a fork truck! He threw it in gear, built up some speed, and then rammed the large forks right through the single brick wall! He bounced in his seat, shook his head and then got down from the fork truck while brushing dust off his pants. Seeing the hole the forks made through the wall, he smiled proudly and said, "Door here! We may now begin the kaizen event."

Now, even I know this was a crazy thing to do. There could be 440 volts surging through that wall, there could have been a water line, or more importantly, there could have been a person on the other side of that wall! Somehow, we must develop a safe **Bias for Action**, or else the resistance to change in any organization will keep you from making any positive changes. Again, please don't do anything dangerous in your factory. But, **DO** have a Bias for Action.

Note: The molding company transformed their operations successfully over the next few years. Eventually, they discovered a better big-picture planning tool called Value Stream Mapping. In their first Value Stream Mapping event, the workers and leaders came up with a process where the molding machine operator dumped the small rubber parts directly into a miniaturized version of the large deflashing machine. It looked like an old child's toy called a rock tumbler. The molding company didn't need the door anymore. All operations, including packing, were now going to be completed right at the molding machine!

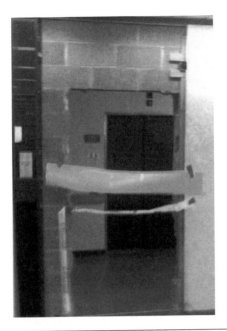

Figure 3.1   A new doorway, but not *that* door.

- OK, that was crazy. But, how can you develop a Bias for Action in your leaders, in all of your workers?
- When was the last time a worker brought up an idea for improvement? How quickly was it implemented? Why?
- The next time you get an opportunity, implement a worker's idea for improvement immediately. How might that encourage more ideas?
- Since the molding company didn't need the door after their first Value Stream Mapping event, was it a good idea to make a door between the two old departments? Why or why not?
- Hint: Toyota will often make a change even for morale reasons alone. They attempt to implement 90% of their employees' ideas or more.

(Triangle Kogyo) Kanban Is Evil/Kanban Is My Badge of Shame (14)

"Continuous improvement is not about the things you do well—that's work. Continuous improvement is about removing the things that get in the way of your work."

Author Unknown

In my evening study of all-things-Toyota, I never lost my desire to learn more about the green golf ball—the pull system—and kanban. I was sure it was *the key* to an effective production system. How do you calculate the number of them? How do you print them? How do you maintain the proper number of them? I had so many questions.

With a few chips in my pocket from problem solving (see My First Lean Tool from Toyota, Chapter 5), I entered Bo's room and knocked.

He grunted.

I entered and said with as much enthusiasm as I could, "Shimono-san, please teach me kanban."

I learned the enthusiasm part from watching the advisors work with Bo.

He said, "I will not!"

I showed Bo a kanban card, "Shimono-san, please. It is the key to the production system I am to improve!"

Bo grunted and said, "Give me that card" (a ratty, creased card).

He took it from me and pronounced, "Kanban is evil," tearing it in half!

Dumbfounded, I picked up the pieces on the ground. Bo had once told me that each kanban card was worth $100 and that the system would shut down somewhere if even one was lost. I said, "Evil? I thought it was the key to our production system?"

Bo pointed to the number in the corner of the card.

He said, "What is that number?"

I said, "35" (35 pieces per container).

He said, "No! What does that number represent?"

I said, "Parts."

He said, "Wrong. What does that number represent?"

I thought, and then said, "Inventory?"

He said, "Correct. And inventory is…"

Waste! Or, evil!

As the light bulb clicked in my head, he said, "Until you remove *evil* from my factory, I will NOT teach you kanban!"

I responded, "How?"

He said, "You will pull cards one at a time from each kanban route and watch the system closely. And, if you shut down my lines, I will kill you!"

I left the room with a torn kanban card, wondering exactly how to remove kanban cards and keep the system from running out.

Actually, it was quite easy. I pulled out cards and watched the system closely. Just before the system ran out of inventory (usually at the customer end of the loop), we would drop in a container of parts with a kanban card one at a time until the system ran smoothly. Then, we would try to solve the problem that was causing the need for more inventory (usually downtime, large batches or quality problems). Then, we tried to pull out one card again the next day after a solution was in place. This sort of physical simulation was quite fun—except for Bo's threat of death for shutdowns. And, it was much more accurate than my previous computer-based, discrete-event computer simulations. It gave us some instant gratification and feedback on ways to lower inventory.

At the end of several weeks of tinkering with kanban card levels, I had a stack of about 50 kanban cards. Not bad, since our system was already pretty lean. I walked into Bo's office with a big smile and a stack of kanban cards held proudly forward.

With gusto, I pronounced, "Shimono-san, I have removed *evil* from your factory! *Now*, please teach me kanban!"

Bo quickly hid the smile forming.

He gruffly said, "Give me a card!"

He summarily tore it (again) stating, "Kanban is your badge of shame!"

I was stunned. I picked up the card and asked, "How can this card be *my* badge of shame! I've only been here for a few months."

Bo said, "This card is for pull system. This card is proof that you are not smart enough to get operation to *one-piece flow*. Kanban is weak third choice! Goal is always one-piece flow! Until you learn to get to one-piece flow, I will **not** teach you kanban system!"

I never did learn everything about the kanban system from Bo. But, I did learn a great lesson about seeking one-piece flow as the *primary* option. Many times, I have seen operations that clearly should have been set up as one-piece flow cells. They remained unlinked because they printed hundreds of kanban cards instead. Bo was right. Kanban is a weak third (or fifth) choice.

Shimono-san's Top Five Priorities for Flow and Pull

1. One-piece flow
2. One-piece flow
3. More attempts at flow
4. Simple pull system like an outlined WIP area (kanban square) or two bins
5. Kanban cards (I lost the first one I touched at Triangle Kogyo)

Figure 3.2   Bo Shimono's Top Five Priorities for Flow and Pull

- Have you ever tried to lower inventory levels (raw materials, WIP or finished goods)? If so, what happened?
- If a person at your facility lowered inventory so low that it temporarily affected production, what would happen? Why?
- What is the best way for you and your team members to discover how to lower inventory?
- If you use kanban in some operations, do you feel that you might have *too many* cards? What can you do about it?
- In some operations, is it possible to apply kanban when one-piece flow should have been applied? Do you think this has been done in some of your operations? What can you do about it?

How Many Kanban Cards? (15)

"If management is not removing the obstacle, they are the obstacle!"

Author Unknown

One nickname I earned at Admiral Engines was "Calculation Man." My Operations Research background was ideal for discrete-event simulation and early applications of Factory Physics.* But, reliance on calculations made at one's desk was not Bo's way.

One day, I found an equation for calculating the number of kanban cards needed in any loop. After researching sources of variation that go into the safety (fudge) factor, I quickly worked up my hypothesis for Bo's new cell. Here was my big chance to show him my quick wit and ability to learn.

After stating my purpose and showing Bo my calculations, I concluded, "We will need 24 kanban cards in this loop."

Bo frowned.

He said, "You will start with 13 kanban cards."

What?! I started to restate my case, but Bo was done.

When in doubt (as long as you have a trusted sensei), do what you're told. I printed 13 cards. A few weeks later, we watched on all shifts as this new cell started up. We were short by two kanban cards, which we quickly inserted.

- How did Bo know that 13 (or just higher) was the correct answer?
- If we had started up the new cell with 24 kanban cards, do you believe there would be any positive pressure to reduce them further?

(continued on next page)

---

* Hopp, Wallace and Mark Spearman, Factory Physics, Second Edition, New York: McGraw-Hill, 2001.

- What goes into a safety (fudge) factor? Does the word *variation* accurately describe some of these? What would W. Edwards Deming, father of the variation reduction movement and early PDCA promoter, say about variation?
- What is your experience guessing the number of kanban cards in advance? What is the correct number?

Note: Once the system was stabilized, Bo always gave me the *removing evil* challenge, "One less than today Steve-san. And, do NOT shut my lines down!"

Paradoxes of TPS: Keep Less There (16)

"Without changing our patterns of thought, we will not be able to solve the problems that we created with our current patterns of thought."

Albert Einstein

TPS or lean has many seeming paradoxes. One of them has to do with the amount of inventory stored line-side for easy use by the operators. One day, we ran out of inventory again on the seat assembly line.

I asked Bo, "Shimono-san, may we please keep additional inventory at the seat track cell? We keep running out!" Bo quickly retorted, "You will keep LESS inventory at the seat track cell!" I argued a bit, futilely. No way. But, when in doubt, try it out.

The confused production and materials workers stared as we repacked all internal inventories into smaller containers. We also shortened the racks that held the containers in front of each operator (flow racks) so that only about half of the current inventory could be stored at the cell. We watched the cell closely over the next few shifts. No material outages. I was pretty sure that was because we were watching. I told the Team Leader and Materials Supervisors to call me when it ran out. After two weeks, I was forced to admit that it had worked.

I asked Bo, "Why did removing inventory reduce material outages?"

Bo grunted.

After a long while, he asked, "What did the *workers* do differently with less inventory?"

I surmised, "They were more diligent and more communicative with the material handlers."

Bo asked, "What did the *material handlers* do differently with less inventory?"

I thought for a moment. Then, a light bulb went on in my head. I blurted, "They were forced to visit each station more frequently." Bo almost smiled.

Victory!

After a moment, Bo said, "Now, go help the material handlers reduce the waste in their routes. You can't just shift the waste to them."

Darn. More work. When will I ever be done with continuous improvement?

- How about you? Do you have outages? Have you tried keeping less inventory with more communication, operator diligence AND more frequent deliveries?
- What will you do if material handlers say they are too busy to visit each station more frequently?
- Just try it. Cut your inventory levels near the operators, fix any immediate delivery problems and then lower it again. At some point, you can cut it too low. But, it is a great exercise proving a key paradox of TPS.

A Rolling Stone Gathers No Moss (21)

"You have to manage a system. The system doesn't manage itself."

W. Edwards Deming

Bo Shimono had a way with words. He also understood American English idioms and analogies. When we undertook a layout exercise, it appeared that we were moving some support departments with little financial justification. Bo said, "A rolling stone gathers no moss." No arguing. We were to move the departments.

One of the departments had not been moved since starting up the factory. They accumulated a lot of junk. On top of that, in an attempt to be "self-sufficient," they had created their own storage areas, printer/copier room, training rooms and even some large walled offices while the rest of us were crammed into an open bullpen area. If a rolling stone gathers no moss, then the opposite is also true. A stone that does not roll gathers a lot of *moss*.

Results: Floor space was greatly reduced. Some inventory was eliminated, and some was put back into supply for the global good of Triangle. But, the bigger result was placing support workers closer to the internal customers they served. This freed up space for more value-added work.

- Describe some of the *moss* that gathers in a department or area?
- How about you? How long has it been since each department or key process area has been moved? The goal is not to move. The goal is to lose the moss. You would be surprised how many times your employees walk around the moss every day.
- Changing the layout costs money. What types of gains or benefits are there to moving things into some sort of improved flow order? Have you tried that with business processes too? Try an administrative worker cell, then reduce the handoffs and try it again.

4

PEOPLE PILLAR PRINCIPLES AND STORIES

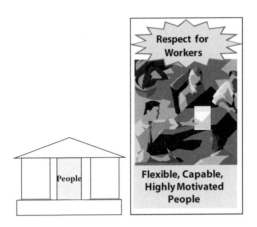

In Toyota's only book published on their revolutionary production system, they state,

> Personal creativity and innovation are recurring themes in the story of the Toyota Production System... Over the years, hundreds of thousands of employees at Toyota companies and suppliers have contributed tens of millions of ideas for improving their own work. Each of those ideas has led to a real improvement in productivity, quality or working conditions. Most of the improvements have been modest, some have been spectacular. But they all have contributed to the continuing, cumulative progress that drives the Toyota Production System.*

Toyota Production System (TPS) principles always show respect for people. If the change does not reinforce respect, it should not be

---

* The Toyota Production System, (Handbook) International Public Affairs Division and Operations Management Consulting Division, Toyota Motor Corporation, 1992.

implemented. This is not to say TPS is easy on people or leaders that are faint of heart. TPS always assumes the very best about people. In poka-yoke, or error proofing, the worker is never blamed. Leaders teach the error-proofing principles to the best workers and ask them to help install true error-preventing solutions so that all employees can do error-free work. They believe that it is management's job to create a system by which people can work error free. It is this uncompromising respect for associates that makes up the center pillar.

All tools and principles must be known and used by the people that do the work every day. Not some separate corporate TPS or CPI team who swoop in for a part-time project. The simple tools must be mastered by the actual workers in their workplaces. That is why people are the most important asset in TPS. That is also the reason why Toyota invests deeply in their people. It also makes good financial sense.

The Power of Ideas

TPS and Toyota's unique way of thinking have given the company a competitive advantage on a global scale. Toyota is known for its company-wide culture of employee involvement and empowerment. In an average year, employees submit approximately 700,000 improvement ideas. Each of these ideas saves money. And most importantly, over 99% of the ideas are implemented.*

Suggestion is not the proper term for Toyota's idea generation system. Toyota's respect for their workers is so great that acceptance of their ideas is a given. Four characteristics mark Toyota's idea generation system:

1. Idea acceptance is a *given*.
2. Pay for suggestions (but don't pay a fortune).
3. Coaches help ideas grow (supervisors review ideas, give direction and hints and help ideas succeed).
4. Implement ideas yourself (no bureaucracy for evaluating ideas).†

---

* Miller, Jon, The Suggestion System Is No Suggestion, Third Edition, http://www. gemba.com Gemba Research LLC, 2003.
† Jon Miller, Ibid.

Toyota workers are the engine of continuous improvement. It is their creativity that leads not only to innovation but also to company loyalty and high morale. Workers with a high degree of ownership of their processes produce great products. And, employees don't argue with their own ideas.

A few years ago, I brought a group of generals and high-level civilians from one of our military branches on a tour of Toyota's Georgetown, Kentucky plant. The Toyota leaders spent hours with us after our tour, patiently answering questions. Among the many questions our generals asked, one stuck out in my mind. Upon hearing that Toyota trains a worker for about six weeks to learn a 60-second job, the general blurted, "That's ridiculous! How can you afford to train just one worker that long?" The Toyota leader asked in return, "What do you do?" This general led aircraft repairs. The Toyota leader was more incredulous that the general did not train their workers for each task longer than they did for a single Camry operation. Lives depended on the aircraft repair person's work.

Over time, workforces change. People come and go. Leaders come and go. Do you have a systematic way to ensure that all workers are certified and capable of producing perfect work and products daily? It seems that companies fear training. The number of different tasks in a job is often reduced to menial, small tasks because a new worker can be taught and made to produce quickly. If jobs are broadened and made more meaningful, it takes longer to ramp up a person's skills. Toyota's goal is that every worker knows every station for a seven- or eight-person work team. The principle is called developing a culture of flexible, capable, highly motivated people. How are you doing in that one?*

Every one of the 20 or so TPS principles and tools needs to be known, used and mastered by the people that do the work every day. It is not enough to have a centralized group of a few TPS-trained people. It is surely not enough to have a small staff with colored belts that

---

* A great resource for learning just how Toyota develops and maintains a culture of flexible employees is Liker, Jeffrey K. and David P. Meier, Toyota Talent, New York: McGraw-Hill, 2007.

swoop in to analyze and then tell you what to do. When the "experts" leave, the workers not only need to make the changes stick, they need to make additional improvements daily. How will they do this unless they have been taught the tools and principles deeply?

A3 Problem Solving

One TPS tool that is used to develop employees and leaders is the A3 problem-solving tool and coaching process. An A3 is simply a single-page format (usually an 11 × 17 inch or European A3-sized paper) for planning and solving problems. A3 planning began in the 1960s as the quality circle problem-solving format. At Toyota, it evolved to become the standard format for problem solving, proposals, plans and status reviews. What is important is not the format, but the process and thinking behind it.

An A3 and the coaching process that surrounds it can serve as an organizational learning tool. It leads to effective countermeasures and solutions based on facts and data, which fosters understanding and leads to agreement. Workers can be taught to speak in terms of an A3. Leaders should request that workers use this format for problems and situations in their area. It adds structure to the employee development process.

Employee Involvement and Empowerment

In this principle, employees are not just *allowed* to improve their work processes, they must do it. The employees are given the keys to their work areas and told, "We trust you. Make changes in the direction of the goals. We'll celebrate when we reach it together." This is empowerment. Empowerment is not something that an employee is told to do. Empowerment is true ownership of a slice of the company itself, with trust that they'll meet their goals.

Employees are often divided into teams, including supervisors, sometimes called Employee Involvement or just Kaizen teams. They work alongside each other in improvement meetings and on the production line as part of a true team.

There is "buy-in through involvement" in applying TPS to work processesses.

My Most Interesting People Pillar Stories

The stories that follow will give you a better feel for the **People Pillar**. Please read these and reflect on the questions that follow each one.

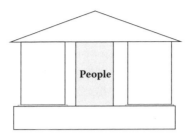

The GUMBI Experiment (5)

"It is not necessary to change. Survival is not mandatory."

W. Edwards Deming

There was a little white lie in one of my previous stories. I said that I worked for Admiral Engines' (AE's) worst factory. We were actually the *second* worst factory in terms of productivity and quality. The worst one by far was the Demot*, California assembly plant. They had frequent wildcat strikes where hundreds of workers just walked out of the factory unannounced.

Demot was also in a gang-ridden area with bars on the windows and doors. Nearby flop houses and belching smokestacks added an eerie appearance to the factory. I heard they had drive-by shootings *inside* the factory! I could imagine a gangster riding inside the cars shooting out the windows as they were being assembled! We heard there were more drugs, guns and prostitutes *inside* the factory than out! AE shut down the factory for a few years, and for that period, I heard that crime statistics actually improved in the area.

Toyota wanted to use AE's transportation system to deliver their cars in the United States. AE had a slick system where cars would roll smoothly from trucks to railcars to trucks. It was a terrific delivery system. In a closed-door meeting, Toyota leaders said, "We will

---

* Another pseudonym.

pay top dollar to rent part of your truck and train delivery system for Toyota cars to our U.S. dealers." AE leaders said, "What else will you give us?" Our leaders were good negotiators. Toyota leaders said, "We will teach you the Toyota Production System (TPS)."

This was the mid-1980s, so very little written knowledge about TPS existed in English. We dismissed any comparison of productivity with the stereotype that Toyota workers in Japan worked a lot harder than our guys. Their *management style* would never work here in the United States and surely would never work in one of our union plants.

The AE leaders asked, "What do we need to do to learn this TPS thing?"

Toyota said, "Just give us a factory in the United States to manage for you."

AE leaders quickly asked, "*Any* factory?"

The arrangements were made in that same meeting to restart and hand over the *beautiful* Demot factory to Toyota! The new facility would be called GUMBI* (Great United Motor Builders, Inc.). Our same union members reported for duty after three long years of layoff. Within one year, they were making the highest productivity, highest quality vehicles in all of the Americas! It was not the workers. There was some sort of magic in Toyota's production management system!

- What if your worst factory was turned around using a different management philosophy? Would leaders embrace it or try to dismiss the results? Why?
- Why do you think AE was so happy to give Toyota the Demot factory? Did AE leaders think Toyota could make the Demot workers productive?

(continued on next page)

---

* Another pseudonym.

> • What do you think the union workers went through those first few months with Toyota leaders running the show? Do you think it helped that they were without work for a while? Do you think any workers quit?
>
> **Note:** If I told you they are still making great cars today, would you believe me? Oh yeah, many of the Detroit operations are gone now. Demot is still going. So are the other Toyota and transplant factories.

(Glass Company) Integrity (8)

"In God we trust. All others bring data."

W. Edwards Deming

In what would be my last month at the glass supplier, I worked for two Vice Presidents—one in Engineering for my "real job" and one in Research as a Total Quality Management Facilitator. Both men had vision and I was proud to work hard for them. The Research guy always drove around in fancy "test" cars like the Corvette. We lovingly called him Dr. Dingdong, mainly for questions he asked early in his career like, "Which way is the glass flowing?" at the start of a mile-long glass production line. In the glass business, we eat our young. We were ruthless when someone questioned centuries-old wisdom.

Our automotive glass plants made some very large parts that hurt our workers' backs, even when two of them worked together to lift them. An engineering team bought and installed two unloading robots as a test. Perfect! The workers thanked us, the company saved money and someone even gave us a President's Award.

The Research Vice President had a great idea. Why don't we buy a large volume of robots and install them at dozens of our plants?

The engineers' eyes widened with excitement, but mainly in fear. That would be a lot of work. We were busy. But, we could do it.

A few months after the robots were purchased, the head of Purchasing asked me in private how much we paid for the two test robots.

I told him.

He whistled and said, "What if I told you Dr. Dingdong paid 10 times that cost for each of those 50 new ones and received some sort of commission in return?"

Ouch. Silence.

I said, "What do we do?"

He said, "Don't say a word. Let me handle it."

He left the company a week later.

Then, I must leave as well. My mind was focused on learning from a real sensei anyway.

Dejected, I started packing boxes. My company had been losing money fast. Human Resources offered employees a few weeks of pay for each year of seniority if they officially retired. I talked to my wife and prayed about what to do. Then, I signed the papers. At the ripe old age of 27, I *retired* from the glass company. On the day I signed the papers, Federal law enforcers marched the Research Vice President and three others out of the building in handcuffs. Little did I know, this was the best thing that could have happened to me. I would have ridden that sad rollercoaster down to insolvency if I stayed.

I interviewed with a Japanese-managed plant near my old hometown. They quickly offered me a job. They also promised that I would have a sensei, or master teacher, to learn their production system in depth. Hot dog! My TPS-style training was about to start. I started at Triangle Kogyo on the first day of my retirement, with a little bit of extra money in my pocket.

[**Note:** The VP was declared not guilty in court because he proved he saved the company a few bucks in each of several unscrupulous transactions.]

- Do "the ends justify the means" in this case? Why or why not?
- When is the right situation to apply advanced technology like robots:
  - When workers are getting hurt?
  - When you can save a few bucks of labor by laying them off?
- Toyota requires suppliers to have "open books" with them concerning expenditures. How might that have helped this glass company?
- What happens in your organization when someone "questions centuries-old wisdom?"

A Good Production Leader Must (11)

"'Mono zukuri wa hito zukuri' is a Toyota saying which means 'making things is about making people'. You cannot separate people development from production system development if you want to succeed in the long run."

Isao Kato, **long-time Toyota Master Trainer (from an interview with Art Smalley, www.artoflean.com, Feb. 2006).**

"In times of change, the *learners* will inherit the earth, while the *knowers* will find themselves beautifully equipped to deal with a world that no longer exists."

Eric Hoffer

By Day 4 at Triangle Kogyo, my original expectations of learning cross-legged at the feet of master sensei Shimono-san had been dashed. I arrived early to put away some personal items—at least the ones that fit in or on the desk. My co-worker, Mitsumi, told me politely but firmly that nothing was allowed on our desks when we leave each evening. The Quality Manager snickered. He was another American who had "jumped in" a few months earlier.

Maybe he was the one standing in *the circle* for two weeks. I made a mental note to get some info from this guy.

Bo had snuck up behind me again. Before he got out the san in Steve-san, I was up and leaning toward the factory. He said, "Steve-san. Today you will start your learning." I stopped. All right! Bo led me through the factory to the back corner of the assembly area. There was a small table with nothing on it but a poorly assembled seat. Bo said, "Take apart. Put back together." Then, he left.

They never taught me how to handle situations like this in engineering school. In fact, they never taught me how to communicate, how to get along with co-workers or leaders, how to work in teams, how to be resourceful, how to solve problems or how to ask questions. What would you do?

I thought, "What would my father or grandfather do?" They would get some tools and just do it. They would probably enjoy this kind of learning. I didn't. Engineers learned from books. I walked back to the Maintenance crib and asked if I could borrow some tools. After some small talk, they rolled out a large, red tool box and said, "Knock yourself out. Just make sure every single tool is back in the box in its outlined spot and roll it back in here before you leave." They loved their tools, just like my father and grandfather.

After hundreds of questions, begging, reading, observing good seats being built, more begging for parts (I ruined several) and three to four very long days, I finished the seat. A bit sloppy, but it actually looked like a seat. I wanted to place this "trophy" by my desk. I found Bo and proudly told him that I had finished the assignment. Bo looked. Frowned. Grunted. He said, "Show up on time tomorrow. You will start your learning." What?

- "Engineers learn from books." How is that working for you or others?
- What is the plan for assimilating a new worker or leader into your organization? Is it written? How fast and effective is your process for getting them ready to succeed?

(continued on next page)

- Are your workers sufficiently trained for their tasks? Are all your workers' abilities tracked in some sort of Training Matrix? Do you reward your workers for their flexibility (ability to do several jobs)? Why or why not?
- How important is product knowledge? Is this important even if your company just does service type work? Why or why not?

A Good Production Leader Must Also... (12)

"It is not the employer who pays the wages. He only handles the money. It is the *product* that pays the wages."

Henry Ford, 1922

"Don't waste time learning the 'tricks of the trade'. Instead, learn the trade."

James Charlton

I mentally reviewed my progress as I drove in that morning. I smiled as I mimicked Shimono-san, and laughed out loud at my own fast-disappearing superiority complex. I walked through the plant first, as would become my daily custom, speaking to as many people as I could. When I arrived at the bullpen area at 6:30 am, Bo was waiting for me. Uh oh. He said, "Time to do the hansei" and walked me to the one closed door conference room in the plant. Bo explained that hansei means reflection.

He queried, "Tell me about your progress thus far."
After the drive in this morning, I was ready for that one.
Bo was impatient.
So, I blurted out my own morning reflections in one long sentence.
Bo frowned.

I asked, "How am I doing?"

Bo said, "Yes, that is the question. How are you doing?"

Hmm.

A question to the sensei always begets *another question*, not an answer.

I tried again, "Am I doing OK?"

He returned, "Would you like to start your learning?"

I said, "Yes, please."

For the next few weeks, I was stationed behind a worker at every station for one *rotation* (about two hours—the time between breaks). It's called a rotation, since all workers rotate to a different station after some time period. One of my advisors explained that rotation prevents some repetitive motion injuries, since different muscle groups are used at different operations. This was further proof of how respect for people drives everything a well-managed company does. I also saw how it helped (forced) us be more flexible and develop our people to do other jobs. Our lines rarely shut down due to the lack of a fully trained worker.

I followed one guy down the line for most of two days, as he was able to work many stations. I observed, wrote, sketched, asked questions (mainly dumb ones) and basically stayed out of the way. For some reason, I was introduced as a temporary employee, not as the new Production Engineering Manager. It didn't matter to me. Other than being on my feet for whole shifts at a time, I was learning. I also learned to take breaks. This was the custom at Triangle, and it was strictly adhered to. Mitsumi would drag me to the break room if I made any attempt to work through the break.

After this period, Bo introduced me to the team as the new Production Engineering Manager. His Japanese advisors smiled and introduced themselves one at a time. I was brought back to my desk and formally introduced to my co-workers. Now, I felt like part of the team. Bo asked, "Would you like to start your learning?" Huh?!

- How about you, "Would you like to start your learning?"
- Is it valuable for people responsible for production to observe every operation before they start leading?
- Can you say, "Our lines rarely shut down due to the lack of a fully trained worker"? Is the level of cross-training tracked for every person? If not, what do you need to do?
- How well did your company handle the workload during July and August (or other) vacation months?
- If you took a vacation this year, was all of your work completed for you before you returned? Could white-collar cross-training and rotation help?

Restarting a Kaizen Team (13)

"The essential question is not, 'How busy are you?' but, 'What are you busy at?'"

Oprah Winfrey

"If you don't know where you are going, you will probably end up somewhere else."

Lawrence J. Peter

Bo showed up more frequently next to me—usually on the shop floor. We just stood together. It wasn't awkward. He seemed OK with me just standing there. Every now and then, someone would ask me to look at something. Bo followed and observed. I drove the conversations in these moments—hoping to get a smile or something out of Bo. He just frowned. He was a hard man. I wrote off his frowns.

One day, Bo said, "Steve-san, you will refocus a kaizen team."
I said, "Great. How do I do that?"
He said, "Yes. How do you do that?"
I begged, "Shimono-san, at least tell me who they are and when they meet?"

Bo pointed to a small group playing ping pong and lounging in the cafeteria.

I walked up to the group and introduced myself.
They ignored me.
So, I name-dropped, "Bo asked me to restart your kaizen team."
More ignore.
I asked, "So, what do you want to work on?"
Shrugs.

At that moment, the Quality Manager rushed into the cafeteria and asked if some workers could help him with a problem-solving exercise on difficult-to-align seat adjusters. The team members shrugged and slowly followed Mike. Hey, problem solving is a form of continuous improvement, or kaizen. I learned two important things that day that carried over to the allotted kaizen time each week:

1. Team members inherently have different input and opinions that are useful in gathering data about a problem and brainstorming possible root causes. They see things differently, and that is a good thing.
2. Using a fishbone diagram to capture the first *round* of root cause brainstorming, then asking "Why" five times (more or less) on each of those causes is a great way to uncover deeper, underlying root causes. The first round of causes brainstormed were usually just *symptoms* of more underlying root causes.

[**Note:** The team did great! All of the team members seemed to know the problem-solving language and methodology. The Quality Manager did a superb job of focusing them during the brainstorming sessions. The team made decisions slowly and carried their ideas to nearly all other production departments telling them what and why. Then, they implemented the solutions very quickly. They also seemed to respect each other's input—even though they argued a lot.]

After a few 30-minute meetings, I reported the limited progress of the problem-solving (kaizen) team to Bo. I told him that

I was not the facilitator or team leader, even though he gave me the assignment. He did not seem to mind. It didn't matter to him that it only focused on problem solving. He said, "Being called a good problem solver is a great compliment. Keep trying."

After several weeks of meetings, trials, sweat, more trials and a string of zero defective seat adjuster days, the team was recognized by Bo in front of all their fellow assembly workers. Before I went home that day, Bo said softly, "Good job Steve-san. Would you like to start your learning?"

- How about your team members? Do they all know the problem-solving language and methodology? If not, what do you need to do?
- Do you have a common form for problem solving? Do all of your employees—even the production workers— know the process?
- Are your teams commissioned to continue meeting until their goals are met?
- Do you involve production workers in root cause problem solving? You would need to free them up to be involved, but what benefit would you get by involving them?
- Do you allow time for team members to do some kaizen (or problem solving)?

You Bad Guy—Reverse Engineering (17)

"Quickly is better."

Bo Shimono

"Quick and crude is better than slow and elegant."

Author Unknown

Working for Bo and his advisors was not easy. In fact, it was the most difficult season of my work career. Yes, I am grateful for gaining a deeper understanding of TPS principles. But, it came at a price that most people do not see.

Each day, I started work at 6 am. Each evening, I went home at 6 pm. As I walked out of the building daily, there stood Bo and the six advisors—usually smoking or chatting. Bo would look at his watch and loudly state, "Steve-san, you bad guy. Why go home so early? Why quit so early?" I felt like yelling, "Because I get paid for 8 hours and work 12!" But, I ducked my head and kept walking. I knew they would only work another hour, and then drink and karaoke until late. The nightly ritual of "You bad guy" continued for many months. I hated it.

One evening, Bo walked up to my desk around 4 pm and said, "Steve-san, you good guy. You go home early to family."

I said, "No. I bad guy, remember?"

He repeated, "No. You good guy. You go home now to family."

I said, "Am I fired?"

He laughed and said, "No, not yet. You good guy."

So, I dropped what I was doing and headed for my car.

As I drove away, this bad guy–good guy turnaround was stuck in my head. I concluded that my Japanese advisors were trying to hide something from me. What was it? Were they going to change something back (again) that I helped implement that day? Were they going to fire the current leaders? I turned my little car around and headed back into the factory.

As I walked toward the office area, I glanced into the Quality Lab window. There they were. All seven of them. They were disassembling some seats—obviously from a competitor. I knew what they were doing! I did reverse engineering at AE. We tore things apart, then placated ourselves about how bad the competitors' parts looked. It took AE weeks to do this sort of exercise, so I turned back around and headed to my car. I'll help them tomorrow.

After a relaxing night with my family, I drove to work the next morning, rounded up some tools and headed toward the Quality Lab. I saw Bo and his key advisors walking out arm-in-arm, saying they were going to get a beer. Now, it was not strange to see or hear this. But, it was 6 am!

I walked into the empty lab. On the floor was a roll of butcher block paper with seat parts cut out and laid out carefully in assembly order. Machines and processes were sketched on the paper under the parts. I did not know what those kanji symbols meant, but I sure knew what the dollar signs meant. In just 12 hours, Bo and team had completely disassembled, processed out and costed out a complete set of seats—containing over 300 parts! I marveled at the speed at which this team *reverse engineered* the seats. I wondered why our guys at AE did not seem to learn as much after weeks of effort using a dozen engineers.

I heard an uncorroborated story a few years later from one of my colleagues. He said that one person named Robert Smith got on the waiting list at Detroit area dealers and ordered the *first* available new vehicle from nearly every U.S. automaker—for full price! He must have a huge garage! A few years after starting up their Ann Arbor design center, Toyota admitted that they were "Robert Smith." Within 24 to 48 hours, Toyota had every single vehicle and subassembly torn down, analyzed, sketched up, analyzed again, processed out and even costed out! The next time you think you are doing reverse engineering, remember Robert Smith.

- Are you doing reverse engineering? What did your team learn on the last tear-down? Maybe you need to start asking tougher questions of this team.
- As you participate in a reverse engineering exercise, note the following:
  - How many parts did the competitor use? How much do they weigh? How did they join the parts together?
  - Can you see any signs of error proofing to reduce assembly errors (true error proofing like one-way assembly and alignment notches, or partial error proofing like labels and color coding)?
  - Do they build in *modules* (versus piece-on-piece assembly)?

(continued on next page)

> - How are their options (e.g., luxury versus economical) included? Are they embedded in modules? How might that ease or error-proof assembly?
>
> **Note:** The competitive response to such rapid reverse engineering is innovation. What percentage of your creative brainpower is your company asking you for and using daily?

Who Should Balance the Lines? (19)

"The most dangerous kind of waste is the waste we do not recognize."

Shigeo Shingo

After the infamous time study incident, I vowed to ask Bo more questions before jumping into assignments. Maybe that is what he wanted me to do anyway.

Bo taught me the fundamentals of Triangle's Standard Work systems, with their two main forms—the Standard Work Diagram for each operator and the Line Balance Chart for the whole cell or line. Bo and his staff taught all of the union workers on the shop floor to use a fairly crude cut-and-paste method to develop Line Balance Charts for various takt times. Takt time is the time in which each unit must be produced to match customer demand. The union workers did their own line balance charts for different manpower levels.

The workers cut out a small rectangle representing each process task from the master Balance Chart Form, where the height of each rectangle represents the time each task took to complete. If we removed some walking or other waste, we would just cut the rectangle down some more. The workers rebalanced the main assembly line based on major break points in volume change by our customer Mazda. We needed to add or reduce workers on the main assembly line so that we always had *equal productivity at any volume* (e.g., the standard manning was 12 workers; but when Mazda dropped below 820 cars per day, the total man-power needed on the main assembly line would be 11; when the demand exceeded 940, we needed 13 workers). It was like a puzzle.

The resulting pasted-up bar charts looked a little rough, but I was impressed that the union workers had all of these bar charts on file for various mannings and volumes. It also made the reassignments on the shop floor much easier. Instead of relying on me or my supervisors to make someone walk to another job, the system was clear to all. When Mazda's demand for the day was posted early in the morning, we sometimes heard groans. But, the workers leaving or coming to the line knew their roles, and the core assembly team started moving materials and equipment around slightly to better support the workers' need. It was a very flexible system.

- How flexible is your system?
- Do your workers help balance the lines? If not, why not?
- Do your workers know the takt time breaks in volume and resulting manning?
- Do you change your manning as your volume fluctuates?
- What is your plan if your customer falls below that volume (e.g., where do the unnecessary workers go)? Can you quickly move the assembly equipment closer to (or farther away from) workers to accommodate this?

Note: Toyota tries to level this demand at least a few months ahead. For our JCI plant in Ontario, Toyota typically gave us 30 days notice that they were going to speed up their main assembly line (e.g., reduce takt time from 60 sec to 55 sec). They asked us all to kaizen out five seconds per station rather than hire additional people—in case their volume does not remain that high. Don't you wish your customer did that?

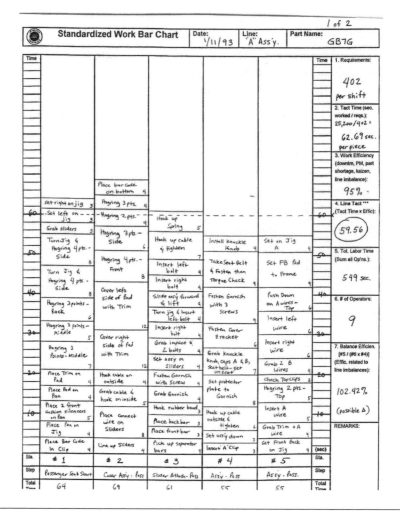

Figure 4.1 Line Balance Chart (rewritten for clarity).

80% Carrot, Steve-san (22)

"Reward those who do, train those who can't, replace those who won't."

Henn's Creed

Here is another of Bo's thought-provoking idioms or analogies. After seeing me express my (AE-learned) punitive management style, Bo marched me into his room. Bo said, "80% carrot, Steve-san." Then, he dismissed me. Was he talking about lunch? Could he be talking about the 80/20 rule? He probably meant that I was doling out 80% stick to my supervisors and workers. After some reflection, I realized he was right. One management expert said to find people doing things right and then recognize and reward them. Boy, that was much more fun.

- Describe some management actions that are *sticks*. Now, think of some *carrots*. Isn't it easier to think of carrots than sticks? Why then do we dole out more *sticks* during a typical work day?
- Are there some team-based, non-financial carrots that you could use to reward teams that are applying TPS principles well?

You Have Lumpy Rug, Steve-san (23)

"We are too busy mopping the floor to turn off the faucet."

Author Unknown

Here is yet another of Bo's idioms and analogies. Bo liked to watch his managers interact with people on the factory floor, as well as their supervisors and peers. He always seemed to be watching, especially if an argument ensued. He never intervened. He just frowned. Typically, we hashed things out American-style. We just ignored them.

After another such episode, Bo again marched me into his room. Bo said, "You have lumpy rug, Steve-san." Then, he dismissed me. Was he talking about my desk area? Could he be talking about the old analogy of sweeping problems under the rug? Again after some reflection, I realized he was right again. I asked for some help when dealing with the Japanese advisors, as well as unruly team members. This style worked better.

- Do you sometimes sweep problems under the proverbial rug? Why or why not?
- If you sought to find the root cause of every problem that hit you during the day and solve it, how long would that take the first few days? How long would that take after a few months of root cause successes?

(continued on next page)

- How about your supervisors? Do they go home before accounting for *all* parts, downtime, changeover and slow operations?
- Do you have a lumpy rug? How can you stop this trend?

Reengineering or Cutting Bone (36)

"The definition of insanity is doing the same thing over and over again and expecting different results."

Albert Einstein

I am sorry to report that a re-engineering assignment dots my career. A short-run book manufacturer brought me in to reengineer some business processes. Their production processes were broken, but I guess they thought the business processes were worse. So, we started there.

Re-engineering offers some great tools for dissecting an organization into its key processes. And, cross-functional process (swim lane) mapping highlights handoffs and gives team members a good big-picture view. But, the implementation methods that I was taught could be described as slash-and-burn. One example of this was cutting the Accounts Payable group at Ford after seeing just a handful doing the same function at their new partner Mazda.

So, they cut and cut our first few key business processes. A few weeks later, we hired back some of those people as contractors at twice their old costs. (This might have been Error #751 or so of the 1,000 or more mistakes I've made.)

- What would happen if only the "fat" was cut out during a re-engineering study? Would the results in the above example have changed?

(continued on next page)

- What is good about "dissecting an organization down into its key processes"?
- What would happen if only the "fat" was cut out during a re-engineering study? Would the results in the above example have changed?
- What is good about "dissecting an organization down into its key processes"?
- There is truth in learning from small and large mistakes. In fact, can you really learn without making some mistakes?

What Percentage of Your Brain Does Your Company Use? (43)

"People don't go to Toyota to *work*, they go there to *think*."

Taiichi Ohno

Because I have asked the question several times already, I better let you know where it came from. I had the privilege of hearing Dr. Edward Marshall* speak to a team of people about change management. Before he showed even a single PowerPoint slide, Ed asked us, "What percentage of your creative brainpower, in the form of ideas, is your company asking you for and implementing daily?" He asked us to mull this over in our heads, and then he repeated his question. He said, "It is very low, isn't it?" He walked us through some of his concepts about building trust and true collaborative environments.

My answer was about 5%. Among those that shared, I had the highest number! There is tremendous untapped value in the heads of your employees. They are just waiting for some way to come out.

---

* Marshall, Edward, Building Trust at the Speed of Change, New York: AMACOM, 2000.

- How about you? What percentage? Why do you think this is so low?
- Lean efforts try to bring out these ideas. How?
- How can you tap this enormous desire by your employees to bring forward ideas for improvements? What would happen if you did this?
- If employees have tried unsuccessfully to bring forward ideas for improvement and were roughed up or turned away, how might they respond to a new call for ideas? What would you do in advance to prepare for that possibility?

5

BUILT-IN-QUALITY PILLAR PRINCIPLES AND STORIES

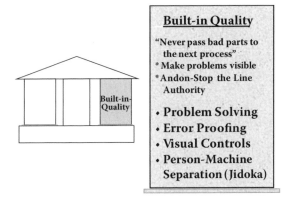

Built-in Quality

"Never pass bad parts to the next process"
* Make problems visible
* Andon-Stop the Line Authority

• **Problem Solving**
• **Error Proofing**
• **Visual Controls**
• **Person-Machine Separation (Jidoka)**

The Built-in-Quality (BIQ) pillar boldly wears its first and primary principle—you must never pass defects or errors to the next process. It holds that quality at the source and immediate root-cause problem solving is necessary to keep the cogs of JIT production spinning freely. They are principles, and they are "musts." In many original Toyota House diagrams, this pillar is labeled Jidoka.

Head Jidoka

Jidoka means that the machine stops itself when a problem occurs. Jidoka is often called "autonomation" or automation with a human element. This concept first perfected at the turn of the century on weaving looms later became a key part of the Toyota Production System (TPS). Jidoka is truly one of the unique inventions of TPS. Originally, jidoka was an invention that allowed Sakichi's automatic loom to stop when a thread was broken. Not only did this invention

111

allow multi-machine handling by workers, the loom had the "intelligence" to detect an error and stop on its own.

There are two main parts to the principle of jidoka:

1. *Separate* human work and time from machine work and time. This principle is based on the belief that humans should do work only humans can do, and machines should do the work of machines.
2. Give a machine the *intelligence* to *stop* when a defect is produced. Sensors are built into machines so that the first defect is detected and the machine is stopped automatically from producing any more. Workers are then alerted and problem solving begins.

This idea of "detect errors and stop" is extended to manual operations such as assembly by empowering workers to *stop the line* when they detect a problem—usually by pulling an andon cord or signal. This is one of the ways that TPS succeeds in building quality into the process by removing defects at the source. Many people mistakenly feel that jidoka is automation for automation's sake. A Toyota worker is taught to automate only after the manual process has been thoroughly studied and standardized so that money is not spent to *automate waste*. The concept of in-station quality control is also included in the jidoka principle.

No discussion of jidoka is complete without some reference to the visionary founder of Toyoda's empire, Sakichi. The son of a poor carpenter, Toyoda is referred to as the "King of Japanese Inventors." His most famous invention was the automatic power loom in which he implemented the principle of jidoka.

Sakichi Toyoda was passionate about jidoka for another important reason. He felt that workers should not be forced to stare at a machine that is automatically working. He likened it to slavery. He felt that the only proof that a worker is *not* enslaved to a machine was if they could walk away from it. This led to incredible productivity gains at Toyoda Loom Works and then later throughout Toyota. But, the principle has its roots in another Toyota pillar and principle—respect for workers.

The BIQ pillar contains some unique and powerful principles. As stated previously, in housing construction, you need to put up the walls pretty much concurrently. This is also true for the Just-in-Time (JIT) and BIQ pillars. As you start going faster, you *will* make errors unless counteractions are taken. They are dependent on each other, rather than in conflict.

Here is a brief analysis of the dependence of the three pillars on each other. True zero-defect systems require one-by-one production and rapid feedback on quality. JIT systems provide this. The workers manage the system moment to moment. Their brains are needed to make TPS work. Sampling inspections do not check all products. And, a production line or cell must stop when an error occurs. The 100% in-station checks of error proofing and problem solving are required so that a JIT production system is not *brittle* due to quality problems. The dependency of JIT, People and BIQ pillars on each other shows how interlaced the principles are.

The BIQ pillar also includes some of the most powerful principles in the arsenal of techniques. The timeless, unchanging principles are:

- Never pass a bad part or error to the next process—Design work so that a product or service cannot leave the station if it is defective. This is also referred to as in-station quality or quality at the source.
- Make problems visible—Make all abnormal situations stick out like a sore thumb; cause the responder to find problems and fix them.
- Andon and *Stop the Line Authority*—Everyone must signal for help when an error or abnormality *might* occur; this is Toyota's number one quality principle.

Some additional tools and concepts are jidoka, solving quality problems, error proofing (using poka-yokes) to prevent defects from occurring, signaling for and receiving help immediately and implementing visual controls to makes problems visible.

Never Pass Defects

Never pass defects is a principle that is well-named. It means just what it says. If a defect is passed to the next station in a cell, it either shuts the system down or (worse) gets more value added at succeeding stations when the product is already defective. When possible, all conditions for zero defects should be checked *before* the machine is engaged. In this way, an error cannot be made.

This principle is sometimes called *quality at the source*. Quality at the source means that defects are identified and corrected as soon as they occur, or at the source, where it is least costly.

When **problems are made visible**, abnormalities stick out like a sore thumb. Systems can be built that create a situation that workers and leaders *must* respond to. How do your workers and leaders become aware of problems? Is it usually too late (e.g., a defect is already made)? Measurement systems can make leaders aware of problems after the fact. But, better systems use the five senses of well-trained workers to stop the process *before* errors or problems can occur. This is yet another way that the BIQ pillar relies on people.

Stop the Line Authority

Stop the Line Authority is Toyota's number one quality principle. But, it is not named well. When a worker pulls an andon cord or signals for help, the line *does not* stop, at least not right away. The worker *does* get an extra set of eyes immediately to evaluate the problem, and the two workers try to fix it together. The person responding to the signal for help is usually a team leader or supervisor, but it can be anyone in an organization. Only if they cannot fix it together in a reasonable amount of time will the line stop so that they can remove the "sick" assembly off-line for extra attention.

When this principle is in place, everyone must stop the line when an error or abnormality might occur. In your operations, *can* your workers stop the line or process if they think an error might occur? Even if they could, as a matter of everyday practice, *do* they? Why not? What are the reasons why a person would hesitate to stop? Some reasons could be fear of punishment (by supervisors), peer pressure, apathy (usually because their concerns are ignored) or a worker fears they would be given more work to solve a problem if they bring it up.

Andon

For this reason, Toyota further developed their Stop the Line Authority systems for production in the United States. Because of the importance of these principles, Toyota measures and rewards the use of their andon cord pulls. Toyota graphs the number of "yellow cord" pulls using statistical process control (SPC). If the number of pulls in a shift goes *above* an upper control limit, they stop the entire section of the line to determine root causes and countermeasures. But, if it dips *below* an expected lower control limit, Toyota also stops the line section to determine root causes and countermeasures. What do you think they are concerned about in the latter case? They fear the worker might get indifferent and pass a potential defect to the next station.

A working andon system brings immediate attention to an error or abnormality that a worker senses. It should create an immediate response from an assigned leader (e.g., a supervisor or manager, but someone from maintenance, engineering or a support function could also be assigned). When a worker requests an extra pair of eyes on a potential error situation, it is indicated via an andon or signal light, board or sound. If the assigned leader does not respond in a short amount of time, the andon system automatically "elevates" the signal to a higher level manager until someone does respond and reset the signal at the worker's station.

Problem Solving

Problem solving is a key principle and tool in TPS. It is found in the Foundation, BIQ and even People pillars. All employees must know and use a common process and form for problem solving. Toyota realizes that all costs due to defects and errors will eventually cost them in terms of profits or even lost customers. They require all of their workers and suppliers to understand and do root-cause problem solving. A simple form is used that usually includes a fishbone diagram. Toyota also promotes the "Five Why's," a technique by which they ask why five times (more or less) to get at the root cause.

Problem solving requires going to the actual place that work occurs. It is sometimes referred to as the three G's—gemba, gembutsu, and genjitsu. These roughly translate into "actual place," "actual thing" and "actual situation." All Toyota workers are skilled problem solvers using three G's.

Error Proofing (Poka Yoke)

Error proofing is a tailorable tool that reinforces BIQ and has wide application. Toyota furthered the tool called poka yoke. A *poka* is an inadvertent error. It is not sabotage or just common cause variation. When even your best workers can forget and create an error occasionally, it is likely an inadvertent error. *Yokeru* means to avoid or prevent. Thus, a poka-yokeru or poka-yoke is something that is put in place to completely avoid or prevent avoidable errors and defects.

Error-proofing devices help build in quality at each production step. This device may take many shapes and designs. Typical types of poka yoke are physical in nature and *prevent* the machine from cycling. They check all possible input conditions *before* the parts are loaded into the machine. They often include sensors, proximity switches, stencils, light guards and alignment pins. Simple circuitry is used to operate these electrical error-proof devices as they should be of low cost and simple design. The goal is to prevent defects before they occur. The goal is zero defects.

My Most Interesting BIQ Stories

The stories that follow will give you a better feel for the **BIQ pillar** tools. Please read these and reflect on the questions that follow each one.

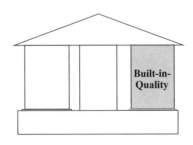

Built-in-
Quality

Standard Work: A Next-Step Quality Tool (20)

"If the student hasn't learned, the teacher hasn't taught."

Toyota Saying

"Improvement usually means doing something that we have never done before."

Shigeo Shingo

Even though standardized work was in the foundation, it is also a powerful BIQ principle. At Triangle Kogyo, I started to monkey with ways to get more work out of our employees. Bo watched me do my own line rebalance sheets for a while, then called me into his office.

He said, "Steve-san. Standard Work—is it a Quality or Productivity Tool?"

I was trained by Admiral Engines.

I confidently stated, "We use stopwatches. We collect time observations. Therefore, it's a productivity tool."

Bo said, "No. At Triangle, Standard Work is our number one Quality Tool."

I returned, "No, Bo. We can get more productivity out of our workers—it's definitely a productivity tool!"

Bo shook his head and frowned.

He said, "It is a *Quality* tool. And for your punishment... You will do Standard Work with the team members every day for the next two weeks using no time observations—no stopwatch, no wristwatch."

I said, "Then what do you want me to do?"

Bo said, "You will work with the team members to seek out the very best sequence and method. You will identify *Key Points* of finesse and assembly techniques that the very best workers use. Then, you will report back to me."

I had a blast. For two weeks, I really focused on the workers and their ideas. Once the workers understood that I wanted to capture their very best tricks-of-the-trade, they enjoyed showing me the *many* subtle ways they built great seats. The best workers had discovered Key Points for every fourth or fifth process task, where they were clearly thinking or doing something different and better than the other, less experienced workers. We wrote these down in the column called Key Points (duh) and showed the other team members. Immediately, the quality improved on the lines where we focused on Key Points. It was small, but you could see it on the daily quality tracking charts. Strangely enough, we also saw small productivity improvements as well. After two weeks, I returned to Bo's office.

I asked, "Shimono-san, how did you know detailed standards would immediately improve quality?"

Bo walked to the board and drew a fishbone or Ishikawa diagram.

He labeled the four main bones of the fish *Man, Machines, Materials* and *Methods* clockwise from top left.

He asked, "Steve-san, can you solve machine and material problems?"

I said, "Yes. Usually."

He asked, "Can you solve problems where the various workers are using different methods and techniques on different shifts and rotations?"

I thought for a moment and then said, "No."

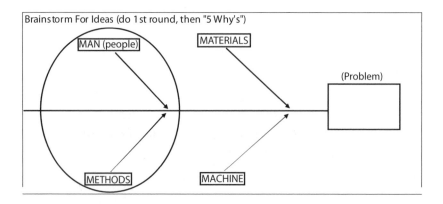

Figure 5.1   Fishbone or Ishikawa diagram; good Standard Work can virtually eliminate potential causes in the circled categories—possibly up to half of them.

Bo crossed out the entire left side of the fishbone (man and methods).

He said, "At Triangle, Standard Work seeks to remove up to half of the possible root causes of our problems. Do you see?"

I saw. Maybe for the first time.

He concluded, "Standard Work is our number one quality tool because it eliminates variation between people and methods. Now, teach this lesson again to our team members lest they forget."

- Review a fishbone diagram from a quality problem, or develop one for your next quality problem. You hit them every day—just pick one. If you categorized possible root causes under Man or Methods, how many of these potential causes could be eliminated by developing and enforcing Standard Work?
- Has a *Standard Work* document been developed and posted for every operator?
- Is it being followed? How do you know?
- Are improvements and updates being made to each document? If not, do you think workers are able to find better methods? Are you asking them to?
- Are Key Points documented for tasks that require finesse, quality, safety or description of better assembly techniques?

My First Lean Tool from Toyota (28)

"When solving problems, dig at the roots instead of just hacking at the leaves."

Anthony J. D'Angelo, **The College Blue Book**

"Don't water your weeds."

Harvey MacKay

After the "You Will Fail" meeting, it was time for my next meeting with senseis Hiro and Phil. As I sat down in front of Hiro and

Phil, I introduced my desire to document a Johnson Control, Inc. (JCI) Manufacturing System, just like TPS. I asked, "What cool TPS tool should I learn first?"

Hiro and Phil seemed pleased and answered in unison, "problem solving."

Problem solving! I better tell them who I am. I said, "Look. I am a *Certified* 8D, Is-Is Not, Train-the-Trainer Problem Solving Instructor. Why do *I* need to learn problem solving?"

After exchanging troubled glances with Hiro, Phil drew the Sea of Inventory (see Figure 5.2 below). He lowered the water level (inventory or lead time) and showed how rocks (problems) that were just beneath the surface of the water now stuck out above the surface. Phil said, "Steve, Toyota believes that we will just *cover up* all problems or *rocks* we hit while implementing TPS systems. There are only two choices when we uncover a rock: we either cover it back up (with inventory or lead time), or we solve the problem—find the root cause and kill it." Phil continued, "Problems are like weeds; unless you get the root, it will grow back." I heard that somewhere before.

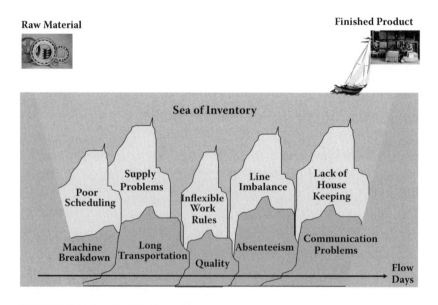

Figure 5.2   Sea of Inventory.

I said, "OK, my plane leaves today at 3 pm. How long will it take me to learn your problem solving system?"

Having fun with my phrase *certified*, Hiro said, "You must be very good, Mr. *Certified* Problem Solving Super Teacher. Because you are already *Certified*, it should take only six to seven... months."

MONTHS! I told our Vice President that I would master the known Toyota tools, document them in our *JCI Manufacturing System* and then spread them to the dozens of other plants—all in one year!

I was a quick learner. After *eight* long months, I had led Phil's teams through enough problem-solving successes to be deemed worthy of the next step by Hiro. Toyota uses a Fishbone Diagram for their root cause "engine." But, they also ask *why* five times or more on each "first round" potential cause (see Figure 5.3). I was told that this team-based, practical problem-solving brainstorming process works about 80% of the time. The other 20% were engineering or leadership solutions, but the quick system worked great in teams.

Two funny notes on the JCI sample Problem-Solving Form above. As soon as we discovered a problem, Toyota required us to get temporary countermeasures in place immediately, create a team, fill out the form through the Action Plan (for the primary contributor) and then fax it to them—all within 45 minutes!

I faxed them the form you see above. As soon as I returned from the fax machine, I received a call from the Toyota Supplier Quality Engineer.

He said, "Steve-san. Goal is always Zero Defects."

My goal line showed a reduction down to two defective Camry arm rests per week. I said, "Look, we had 22 bad arm rests out of 5,000 seat sets last week. So, two is pretty good, right? Two is good?"

Silence.

The Toyota Supplier Quality Engineer said, "OK. You can have two defective arm rests per week. But, YOU must buy both Camrys with defective arm rests every week!"

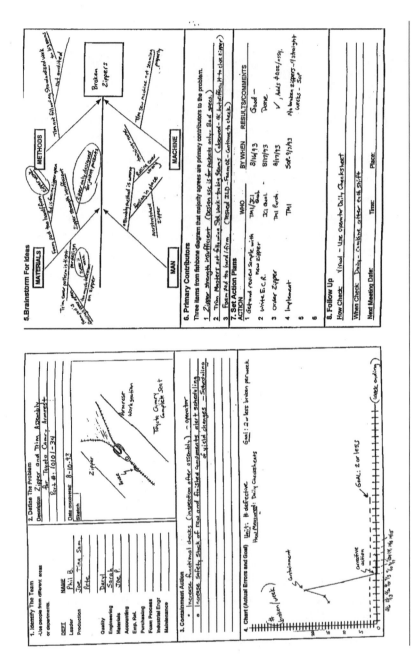

Figure 5.3 Toyota Practical Problem-Solving Example.

I quickly agreed, "Zero Defects sounds good to me. I'll change the form."

A second funny story from this Form concerned our Temporary Countermeasures. I was forced to remove the word *Inspectors* from Box 3 because JCI didn't like that word. But, when I faxed the original form to Toyota, it showed three additional JCI inspectors per shift within our plant before loading the trucks.

The Toyota Supplier Quality Engineer also said in his first call, "Your countermeasure is not *costly* enough!"

I said, "We have three people for both shifts looking for bad arm rests. What more do you want?"

He said, "Six!"

I asked, "Why?"

He said, "If we don't make the Temporary Countermeasures *costly* enough, you will just leave them as part of your *permanent* operating system. You are our key supplier. Eventually, we will bear some of your unnecessary costs and waste. We will not allow that to happen."

I was thinking where six additional inspectors would even stand around our small assembly cell. Six per shift! Well, it worked. The costly temporary measure kept pressure on our team, who quickly found the root cause—a plastic zipper had been changed (it was model year change time, and this was an untested attempt by an engineer to reduce costs). Toyota taught our engineers how to use a reverse fishbone before making a change. You put the possible countermeasure or solution "use cheaper zipper" in the back head of the fish, then brainstorm ways that a change like this might *cause* new problems. It works.

- How about your facility? Do *Temporary* Countermeasures like added inspections become part of your *permanent* operating system because you did not get the root?

(continued on next page)

> - Do all of your team members know a common language (methodology) for root-cause problem solving? It is not enough to have a form. Does your form have a box labeled "Write your Root Cause here"? (If you knew the root cause, you wouldn't need a form.)
> - Do your engineers always use a reverse fishbone diagram when evaluating changes?

Commissioning and Cross-over Shift Meetings (29)

"If you don't have time to do it right, you must have time to do it over."

Author Unknown

I have learned a lot about building problem-solving systems. We always used a tracking chart to show not only how bad the problem was on a particular day or week, but also the trend. We were taught to put a goal on the tracking chart. We would also write "Big Celebration" near the goal line. Toyota required suppliers to meet the goal for six straight weeks before the problem-solving team could celebrate and then stop meeting. It was tough love, but it kept us from stopping too soon.

Toyota *commissions* a team to do problem solving. For them, a problem-solving team is like a military Special Forces team. When a Special Forces team is commissioned, they either accomplish the mission, or they die trying. There is no stopping, getting distracted or quitting. I'm not trying to be morbid. I have tremendous respect for these special warriors.

In the Toyota Practical Problem Solving (Figure 5.3), our team had three straight weeks of zero defects. Three weeks later, there really was a great celebration in the land. It was fun to have a Toyota leader celebrate with us and recognize the team. They also received t-shirts and a sincere thank-you from Phil and the JCI leaders.

One note on meeting times. Logically, our small plant could not *afford* to have 30 or more teams of 6 to 10 people meeting weekly until they achieved six straight weeks of zero defects. But, Toyota demanded this. The alternative was even more costly—really. Toyota recommended that we use a cross-over shift meeting. Teams were to continue to meet until they achieved the goal. Team members were to be freed up from their production tasks to attend.

After the initial meeting (usually for one hour), teams were commissioned to continue meeting daily or weekly (depending on the severity) for 15 minutes at the end of the first shift and 15 minutes at the start of the second shift. They only paid 15 minutes of overtime per team member per week. It reduced overtime. But, it also allowed us to buffer the loss of too many workers on any one shift by using team members from both shifts. That was cool! This also allowed us to quickly implement solutions, because multiple shifts already had buy-in to the ideas being considered. It streamlined communications and empowered team members. As we increased our cross-shift meetings, many of the battles between shifts eased as well.

- Do you commission your teams? Do they have the mind-set of a special forces team? Does management really plan to allow them to continue meeting until they meet the goals?
- When do they meet? Where? Do you use cross-shift meetings?
- Have you had difficulty spreading ideas or standards to other shifts? Were they involved (represented) in the solution?
- How do you celebrate problem-solving successes? Do you share these successful forms with other plants and sites?

Stop the Line Authority (30)

"Simple, clear purpose and principles give rise to complex intelligent behavior. Complex rules and regulations give rise to simple stupid behavior."

Dee Hock

After mastering problem solving and many of the base tools, we actually did develop some one-piece flow cells at JCI. We reinforced them with visual management, and even got into quick changeover (SMED) and total productive maintenance. But, as we started to build faster, we also started making defects faster. We needed to button up quality and do it fast!

It was time for my next meeting with senseis Hiro and Phil. I told them of our changes and the battle to sustain. They smiled and encouraged me. I also told them that we started to make more defects. Smiles disappeared.

I asked, "How does Toyota get great efficiency with perfect quality every day? What is the key *quality tool* that Toyota uses to achieve both at the same time?"

Hiro and Phil looked at each and then said concurrently, "It is *Stop the Line Authority.*"

I said, "No. I am asking for the key quality *tool*. You know, like poka yoke, SPC charts, or jidoka."

They frowned and then repeated, "We told you. It is *Stop the Line Authority.*"

I gave up and started writing as they described this concept to me. Hiro said, "Any one *must* stop the line when an error or abnormality *might* occur. We measure and reward this. The employee uses some sort of andon or help signal, sometimes shown on a hanging display board. This should bring immediate response from a supervisor or manager."

As stated previously, Toyota tracks the number of andon cord pulls every shift on every section of their lines SPC style. If they get too many pulls, they stop the line, gather the workers together and solve the problems (e.g., a rash of bad supplier parts). But, if they get too few pulls, they also stop the line, gather the workers together and solve the problems.

The line or process does not immediately stop; it merely brings another set of eyes right away. They try to fix the problem together while the line still moves. Then, they decide together if the line actually must stop to solve the problem.

The supervisor is trained to say the same thing each time they respond, "Thanks. What do you see?" They assume the worker has caught a problem that could affect the customer. The worker has ultimate authority to keep defects from getting to the customer. It makes you want to buy a Toyota, doesn't it? Healthcare companies are also starting to implement this key principle.

I thought of a hundred reasons why our workers would *not* stop—mainly due to yelling by supervisors like me. We had a long way to go on this one. But, we tried, and tried...

- How about you? *Can* your team members all stop the line?
- As a matter of daily practice, *do* they stop the line? If you are hesitant to say yes, what are some reasons that workers do not alert someone if there *might* be an error? How can you eliminate these reasons?
- Andons do not actually stop the line. What happens when a person pulls the andon cord or calls for help? That's right—someone comes to help. It brings another set of eyes to the abnormality. Who will help immediately at your location?
- When Toyota uses SPC-style tracking and stops the line for too few andon cord pulls, what do you think they are concerned about?

The Expensive, Broken Andon Board (31)

"The road to success is always under construction."

Chinese Proverb

A few years later, a student from one of my Lean courses called me one day to report on his implementation of an andon display board (see Figure 5.4), almost exactly like one of my slides.

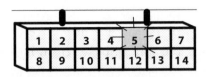

Figure 5.4   Simple Ceiling-mounted Andon Display Board.

He said, "$167,000!"

I said, "Hello?"

He explained that his company needed quality improvements fast, so he immediately returned from class, installed a 14-station andon board and wired it into all work stations on a line.

He said, "The workers pushed their andon button, but no one came!"

I thought to myself, "What did this guy expect—Maytag repairmen rapelling down from the ceiling?"

I said, "OK, how can I help you?"

He said, "It's OK. The andon board system broke the next day."

I said impatiently, "Please get to the point."

He said, "So we went to a discount store and bought some of those big orange bicycle flags—you know, the ones on the long fiberglass pole?"

I said, "Yes, go on."

He said, "When a worker needed help, he raised up the flag."

I said, "Yes, yes, please go on."

He said, "But still no one came!"

Now, I was frustrated. I asked with gritted teeth, "How can I help you?" He said, "I just wanted you to tell this story to your classes whenever you teach this principle. Eventually, we worked out *who* responds and *how*, and then freed them up to do so. That's what we really needed. Please tell your students that they need to work out some kind of supervisor response system first." Touché. I learn from my students every single day. I even learn from their questions.

- How about you? If one of your good employees alerts someone that there *might* be an error, how would they do that?
- If someone pulled an andon cord, would anyone *ever* come? Why or why not?
- Are you encouraging, tracking and rewarding them to do so? How might that be a catalyst to get employees to use the system?
- Starting with just one cell or line, how might you work out *who* responds and *how*, and then free them up to do so? What is stopping you?

Elevating the Andon Signal (32)

"What you see depends on what you thought before you looked."

Eugene Taurman

I have taken leaders of companies on tours through Toyota's operations in Kentucky a few times. On one occasion, a small group watched an assembly line while waiting for the rest of our team. Toyota does not use complicated andon light systems for all signals. They are expensive, as you noted in the last story, and not all responders can see the light board from all locations. You need line of sight for andon boards. So, Toyota uses sound or musical jingles more and more to denote which line or responder is needed.

A line andon jingle started playing over a loudspeaker. The Toyota leader pointed in the direction of one section of the line. A flashing light was also spinning where the problem was happening.

After another minute or two, our leader announced, "Good. You are all going to see this for yourself."

I said, "See what?"

He said, "After three minutes with no one responding, the andon signal will elevate. This typically means the responder's supervisor or manager must now respond."

He said, "Here it comes…"

The fun-sounding jingle turned into an onerous "Duh, duh, duh duh. Duh, duh, duh duh."

Before the next note, a manager with tie askew ran from the office area toward the flashing light. He was the fall-back person for the Team Leader. He seemed to be looking for the Team Leader, who was helping to treat a worker that had fainted. Toyota's andon signals keep elevating until even the Chief Executive Officer must respond. By the way, you don't want that to happen. It is a true 24/7, drop-and-run system that requires a response.

- How about you, have you worked out who would respond? Do they have time?
- What is meant by a "24/7, drop-and-run system that requires a response"? Do you have that yet? If not, what is still needed? What is the next step?
- Have you built elevation responses into your system?
- What is more important, a meeting or a five-minute or less response to the line worker's andon signal?

6
WRAPPING IT ALL UP
(THE "ROOF" OR RESULTS)

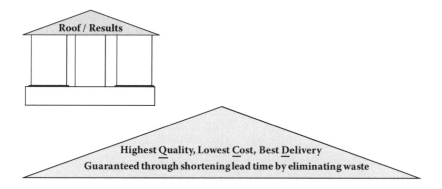

The most important part of the Toyota House is not the solid founda-
tions or principle-laden pillars. Toyota-trained sensei may argue that
the center Respect for People or Culture pillar is foremost. But, this
author believes that the house itself is the key. There is no way to take
a single principle and separate it from the others. The simplicity and
beauty of the house implementation model is apparent.

A critical focus of the Toyota House is the goal. The North Star
is the roof or results. Notice that Toyota Production System (TPS) is
not the goal. Neither are any of the tools. The goal is to build a strong
enterprise that produces goods and services of the best quality, cost
and delivery in the business. Guaranteed.

It is this guarantee that drew me toward TPS and the pursuit of
all things Toyota. I have never heard anyone speak so confidently of
other process improvement methodologies—and I had been through
them all it seemed. The guarantee is this: *if* you build a strong foundation, then build everything just-in-time, while building in quality, all
the while developing a culture of flexible, capable, highly motivated

workers, and you will get the best quality, cost and delivery in the business. Guaranteed!

What Is Success?

Organizations implementing TPS vary widely. The interesting thing is that their successes share a definite pattern of fits and starts, followed by stabilization, only to be upset again with systemic change. The other similarity that marks successful firms is their results. If one looked at a single metric, say productivity, it would appear to increase slowly for a period of time and then take off exponentially.

One also needs to define success. Success, for the purposes of this book, is sustained results a year after the major changes are made. Many metrics can be used for this, but overall end-to-end lead time, inventory turns and defects should be measured, at least.

Measurable improvements always follow system-wide TPS implementation. These results have been validated through industry averages[*] multiple times. The typical results of organizations achieving success through TPS are:

- Direct Labor/Productivity Improved   45–75%
- Cost Reduced   25–55%
- Throughput/Flow Increased   60–90%
- Quality (Defects/Scrap) Reduced   50–90%
- Inventory Reduced   60–90%
- Space Reduced   35–50%
- Lead Time Reduced   50–90%

Sweat and Blood

True TPS implementation and culture change also require hard work. In the words of Taiichi Ohno, "You should submit wisdom to the

---

[*] Womack, James, Arthur Byrne, Orest J. Fiume, Gary S. Kaplan, and John Toussaint, Going Lean in Health Care, Institute for Healthcare Improvement White Paper, 2005.

company. If you don't have any wisdom to contribute, submit sweat! If nothing else, work hard and don't sleep! Or resign!"

How about you? Are you ready to do the hard work of change? The toughest thing to change is human behavior. That is changed one process improvement at a time. Give your people the keys to their own processes. Teach them to remove waste and solve problems. Watch what happens.

My Most Interesting Roof Stories

The stories that follow will give you a better feel for the **Roof** of the Toyota House. Please read these and reflect on the questions that follow each one.

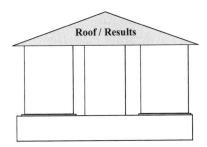

Keiretsu Gone Bad (25)

"Failure is only the opportunity to begin again more intelligently."

Henry Ford

Sometimes even lean companies do not survive. Sweeping change can cripple a company that is stuck in old modes of belief.

A rumor rippled through Triangle that Mazda's Big Three partner just took a majority stake in our customer plant. Breathless, I barged into Bo's office and blurted, "Is it true?" Bo dismissed the other advisors and frowned at me. He said, "Steve-san, do not concern yourself with rumors. We are in a keiretsu—a customer family with loyal suppliers for life. Every Mazda car has a Triangle seat. It has always been that way and will always be that way!" Bo's voice rattled the walls. It was almost as if he was trying to convince himself this would remain true.

I tried my best to inform Bo of American purchasing strategies. They knew nothing of keiretsus. I begged him to press for new contracts and seek other work. But, loyalty is the code in a keiretsu. There is no "other customer." A competing American seat maker was already making the other half of the seat sets at Mazda.

I knew what I needed to do. It took me only three weeks to interview and sign on with America's largest seat maker, Johnson Controls, Inc. (JCI). Within a few months of my leaving, Triangle lost its seating contract. The Big Three partner threw them some small parts for a truck plant down south. Within a year, Triangle shut its doors. Loyal and hard-working Bo Shimono took a plane back to Japan, a defeated man. But, his principles live on through a handful of trainees (still) in training who are humble enough to stand in circles. No regrets.

- Who is your sensei?
- If you have or had a sensei, when was the last time you felt uncomfortable? If it's not a bit painful, you probably don't have a sensei.
- Assignment: Google the phrase Supplier Working Relations Index, find out as much as you can about the topic. If you are an Original Equipment Manufacturer (OEM), self-assess yourself against the 18 areas, and then implement immediate actions to improve. If you are a Supplier, self-assess yourself against the 18 areas, and then implement immediate actions to improve. You have suppliers too. What should your score be?
- What could Bo have done to protect himself against total dependence on one customer?

(Consulting) It Works! (37)

"Waste is a tax on the whole people."

Albert W. Atwood

One of my early lean successes as a TPS or Lean Coach was at a robotic controller assembler. About this time, Mike Rother and John Shook had completed their great how-to book Learning to See, on Value Stream Mapping (VSM) arguably one of the most useful tools in the TPS arsenal. Mike shared with me some of his drawings and notes behind the tool, and I was off.

We used leader-led VSM. The team drew up a great vision and then set out to implement it. We did not use the typical five-day kaizen events, which was a first for me. A strong, active Sponsor and a committed Team Leader kept the team focused on the detailed action plan. Overall, lead time was reduced 97%, Work in Process (WIP) was reduced 50% and productivity doubled, all without layoffs. See the results in the Figure 6.1 below.

Similar results followed on other large projects and companies. I noticed there was a clear relationship between results and setup effort! Active setup in advance of workshops produced great results. Incomplete or rushed setup didn't fare so well. The same thing held true when an active Sponsor and committed Team Leaders were not prepared in advance.

I helped author and professor, Dr. Jeffrey Liker, start up his lean consulting firm, Optiprise. We helped implement TPS principles with companies that thought the principles would not work. If I only had a nickel for every time I heard, "But, we are so different."

	BEFORE	AFTER	% IMPROVEMENT
Floor Space	9820 sq. ft.	3600 sq. ft.	63.3%
Total Cycle Time	247.0 hours	8.2 hours	96.7%
Part Travel	2.1 miles	0.4 miles	81.0%
WIP	—	—	50.0%
Staff	10 workers	5 workers	50.0%

Figure 6.1   Results from Value Stream Improvement Project.

After several great successes, I finally started to feel comfortable with the improvement toolkit. But, I never stopped learning.

> - How about you? Do you feel that you know all you can about the continuous improvement toolkit?
> - How good is your active setup in advance of workshop? Think of your last few change events. For the good ones, was the setup good? Is this a coincidence? Is it possible to develop a bullet-proof setup checklist?
> - Do you implement your value stream action plans part-time over a long period, or do you use only kaizen events or workshops? Is there wisdom in doing some of both?

A "Death Notice" Is a Burning Platform (38)

"If you don't like change, you are going to like irrelevance even less."

General Eric Shinseki

When I agreed to serve as a lean coach for a muffler manufacturer, I was concerned about the superiority complex auto assemblers were infected with. As Dr. Jeff Liker and I spoke and encouraged these companies to try a true Toyota-style approach, they were quick to say, "Been there; done that; got the t-shirt."

This muffler facility was different. There was a tiredness there that you could see. They were tired of layoff cycles. Cut when times were bad and hire back (some) when times are better. High labor costs and distances from customers and key suppliers seemed certain to sink this once proud factory. Corporate leadership sent this facility a "Death Notice" (notice of pending shutdown). They had about six months before shutdown.

I asked the Plant Manager what it would take to stave off the shutdown. He said, "We need $1 million of savings quickly." They had already laid off many of the second- and third-shift workers. He was out of ideas.

The Plant Manager walked me through the facility. Every available space and even the aisles were laden with WIP inventory.

I smiled and said, "Sir, it's time you took out a large withdrawal."
He said, "No use. We have no cash on hand."
I said, "No, wrong bank. You need to take out a large withdrawal from your bank called inventory!"

Within days, we had a VSM plan and vision for flow and pull. Now we needed to sell these major changes to the workers.

In a series of all-employee communication meetings, we demonstrated the power of flow and pull using a simple hands-on simulation. The employees laughed and were skeptical. I can't blame them. Every annual corporate program thus far left them worse off and ended in layoffs. The Plant Manager showed them the plan to remove $1 million of WIP inventory. He said it's this or shut down. After a brief pause, a large man stood up and said, "C'mon guys, let's try this flow and pull stuff."

Within six months, this facility had removed about $1.4 million of WIP and other waste. Within one year, the facility had won their company's *Most Improved Plant Worldwide* award. The plant loses and then gains back their TPS-style focus when a new Plant Manager comes in and tries to avoid involving the workers. Once a workforce is involved heavily in TPS deployment, they will not want to go back to a traditional hands-off, brains-off style of management. This plant is still there. Living things don't want to die.

It sure is easier to implement rapid change when you have no other options. Maybe that was a blessing to this facility.

- Have you taken out a huge withdrawal from your largest bank yet?
- Is it right to fake a Death Notice? If not, what *burning platform* for change can you communicate to convince people that action is needed?
- Are employees desensitized or immune to words of pending doom?
- Why do you think this plant would "lose and then gain back their TPS-style focus"?

When Will I Be Certified? (39)

"Amateurs work until they get it right. Professionals work until they can't get it wrong."

Author Unknown

"When will I be certified as a TPS Facilitator or coach?" Many people have asked me this question, especially in government organizations for some reason. I tell them that TPS principles are *experientially* learned. It is not enough to take 10 days of boring classes and then do one small project.

In practice, it is possible for someone to conduct successful TPS workshops or events after just three steps. In medical training, they say, "See one, do one, teach one." In TPS, we like to see a prospective facilitator be a participant in a workshop, then co-lead a workshop with a trained sensei and then try one on their own. They will still make some mistakes after this, but small mistakes help build the facilitator's true experience and knowledge. There is no better teacher like experience.

George Koenigsaecker* successfully transformed several good American firms using TPS principles. George likes to say that a person is not *certified* as an expert TPS Coach until they have completed 60 or more successful lean projects. George is on to something. My effectiveness with the larger system of tools (not just one or two) was suspect until I reached that point. Even though this level of experience would have sounded high to me in my early years, I believe this to be a true test.

Bo Shimono from Triangle Kogyo never really said I was *certified*. One day, he just stopped yelling at me.

---

* President of Jake Brake, then CEO of HON Industries, now president of Lean Investments, LLC, George relates his experiences in Koenigsaecker, George, Leading the Lean Enterprise Transformation, New York: Productivity Press, 2009.

- Do you desire *certification* as a Lean Coach? Why? Was certification promised by a consultant? What would you need to do?
- How many successful projects have you yourself facilitated or directly led (not just swooped in for the dog-and-pony show)?
- Do you feel that 60 lean projects is too many before someone is ready? Why or why not?
- How many projects or events are needed before you can try one on your own?
- Hint: Bo would say, "Quickly is better."

Body Audits and Layered Audits (40)

"Why not make the work easier and more interesting so that people do not have to sweat?"

Taiichi Ohno

Another successful implementation of TPS principles was at a lens coater. It took an average of 14 days to receive, coat, process, pack and re-ship the lenses. They also had over $1 million of WIP tying up every inch of floor space. Piece of cake.

We quickly identified the bottleneck (Tip: inventory piles up in front of the bottleneck process step). We taped off some simple kanban squares in front of the bottleneck. We dedicated five cart spots—one for each bottleneck stamp/pack station (see Figure 6.2 below). Then, we documented, communicated and enforced this new WIP limit and pull signal. When all spots were full, the operator at the previous step (bake) would open the oven door, but leave the cart laden with ceramic trays and lenses in the oven.

The operators cried, "But then we can't bake any more lenses!"
I said, "Right. You don't need them."
The operators said, "What shall we do?"

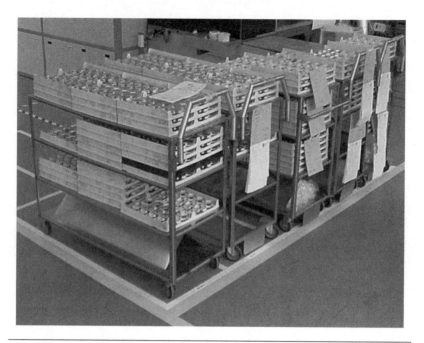

Figure 6.2   Simple kanban squares; *pull* signal.

I said, "Help stamp and pack lenses. If you bake more lenses, they will just sit in the aisles. In order to ship more lenses today, we will fill the WIP-limited 'supermarket' to its maximum, then go help."

Human Nature: The workers had been told to "stay busy" and "just bake lenses" for years. I knew this would be hard for them. I am a crazy coach. Not only did I help them implement the key change, I was also there on Monday standing next to the key changed areas to help keep them from reverting. I call this a "body audit." Physically standing next to key process change does wonders for their enforcement. It also gives the leader a great chance to explain to workers *why* the change was made and which of their fellow workers had come up with the great ideas.

Within 10 minutes of filling the WIP area, one of the bakers came around the corner with a full cart of lenses. He saw me with my arms crossed next to the full WIP limit area, scowled and then backed up to the oven. The Law of Least Effort works against changes like this. If the baker operator can load up a few ovens and then kick them on for an eight-hour bake cycle, he could go to the cafeteria and play some card games.

You need to audit all key changes. And, you need to audit them frequently. Once everyone has passed every audit on every shift for a month, you can back off to weekly audits, then maybe monthly audits. Workers know that we only audit the important things.

Another key concept is a layered audit. Starting with your key changes, have a top leader audit the physical audit sheets that the team leaders or supervisors have completed. They need to audit the audits. Ask tough questions. These layered audits must be done immediately or at least very soon after the team leaders or supervisors document their audits. And, it must be done where the work occurs. You would be surprised how much "pencil whipping" the supervisors do when they are busy. So, don't let them.

It worked! Lead times reduced to two days and over $1 million of inventory was pulled out of working capital within 90 days!

- Think about the things that you audit in your facility. Are those the only important things? Are there others that should be audited?
- Do you audit all key changes? Who does it? Is it documented?
- Do you audit the audits? This does not need to take a long time. Just choose a sample and check them out. What did you see?

The Terrible Lean Success Rates (41)

"An environment where people have to think brings with it wisdom, and this wisdom brings with it kaizen."

Teruyuki Minoura

You may have seen similar statistics. Lean projects are successful only about 10% of the time (see Figure 6.3). This study by the

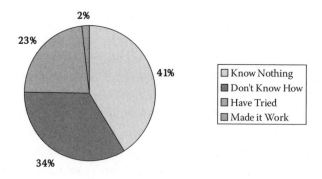

Figure 6.3   From: Association for Manufacturing Excellence (AME).

Association for Manufacturing Excellence (AME) asked company leaders about their usage, success and sustainment of Lean principles. Forty-one percent of those who were called answered something like, "What is Lean; some kind of diet program?" The AME surveyor thanked them and hung up. Another 34% of those called answered that they had heard of Lean-type improvements but did not know how; 23% actually tried to implement the principles and only 2% successfully made the principles work in their operations. Less than 10% of those who knew and tried these principles were successful at achieving the goals.

Success is sustained measurable results after one year, so this is tough to do. My partners and I were certain that our "batting average" was better. We use leader-led VSM and visioning. We teach them great setup, change management and follow-up. There is no way we can be that low!

I called several of the leaders who attended courses I taught. The truth: only a few percent more than the national average succeeded. Rats! I asked some deeper questions. Since VSM was the key tool we were advocating, I asked if they used it in leader-led teams. Only about *half* of them had done Current State Mapping in teams! The others typically said, "I wanted to do VSM in teams—but our leaders would not let me! They thought by paying for me to go to training, I would somehow come back and transform our company without any other person participating in a team or study!" Double rats!

It also seems that the time limit for VSM is three to four months. If you do not use the VSM tool within about three to four months of learning it, you will NEVER use it. It's like learning to fly a plane by reading a book. Eventually, you need to get in a plane and fly.

I dug deeper in my simple phone survey. Only about half of those making a Current State Map (about 25% of the total group) made a Future State Map. Incredible. The conclusion of a current state map is, "We are awful!" Why would anyone stop there! Some of them said they just cherry-picked a few problems from their current state and started implementing. Of course that didn't work! That is like straightening the lounge chairs on the Titanic! The ship is sinking. You'd best learn to swim (get the process to flow).

Even more incredibly, only about half of those making a Future State Map (about 12.5% of the total survey group) kept on going to make a detailed action plan AND *resourced* the plan. By resourced, I mean that management made sure key action items were high on the list of their assignees to do. They re-prioritized their duties but did not hire more resources.

Conclusion: Of the 12.5% or so of the group that completed a detailed action plan and resourced the plan, 80% to 90% of them succeeded! We knew it! It wasn't 90% fail. It was 90% succeed! Just follow the process. Do leader-led VSM with great setup, change management and follow-up. And, follow the process.

- Of all of the improvement efforts you started in the last decade, what percentage succeeded?
- Why does the probability of success improve so much when the action plan is fleshed out and resourced?
- Why is solving small problems in your Current State like straightening the lounge chairs on the Titanic?
- Have you completed a leader-led VSM package? Did you *resource* the main deliverable—the action plan?

7

CLOSING WORDS

Deploying the Toyota Production System (TPS) principles is much akin to dieting. (Perhaps that is why the phrase "lean production" has stuck over the past two decades.) In dieting, the basic concepts are quite simple: eat less, move more, eat healthy food, don't eat late, and so forth. But, dang is it hard to do day after day! Everyone wants the bariatric surgery type of single change. But, that is not how the principles are implemented.

One person described TPS as 1,000 little things done right every single day. That thought scared me at first because I did not believe many organizations have the guts or staying power to accomplish 1,000 of anything. Most firms want a single silver bullet solution. But, that does not exist.

A baseball analogy works well for implementing TPS. If you are the leader of your organization, train and then learn to trust your people to hit a bunch of base hits every day. Don't swing for the fences. The team that gets 1,000 base hits will win every time.

Using a track analogy, continuous improvement is a marathon, not a sprint. Continuous improvement is just that—it's continuous. Teach your employees to identify and reduce waste. Then, allow them to do that every single day.

Which stories stick out the most in your head right now? It is likely that the parts of the Toyota House represented by that story need some work at your facility. Could you write down all of the silly or dumb things I did in these stories? How could you prevent them from being made at your company?

This is also a great time to re-read some of the questions at the end of each story. If you are anything like me, it takes many questions like these to convince a person that change is needed badly. You are not alone if you are struggling to make changes and make them stick.

Here are just a few tips from the book that you might want to consider:

- Stand in circle.
- Find and use a Toyota-trained sensei whose learning curve you can borrow to get you past the bear traps of implementation.
- Train your leaders and middle managers so that they have a common language for TPS and changes that will follow.
- Learn by doing—set up and do some learning projects right away.
- Just do it—start your journey—don't wait.

Acronyms and Some Terms

The acronyms provided in this attachment have been largely derived from a variety of continuous improvement publications and public documents.

12-Gun	A large human-controlled tool that torques down 12 bolts at the same time
5S	Traditional Toyota Production System tool to clean up, organize, and standardize a workplace: Originally five Japanese words starting with the letter S, translated to several combinations of English words. This is one interpretation: • Sort & Scrap (clear out rarely used items) • Straighten (organize and label; a place for everything) • Scrub (clean) • Standardize (make standard the best-known way to do something) • Sustain (audit and reward the previous four items)
80/20	The 80/20 rule; Pareto rule (e.g., 80% of the problems are from only 20% of the possible causes)
8D	*Eight Disciplines* or steps of root-cause problem solving; used by the Big Three automakers
A3	A problem-solving format, process and development tool
AE	Admiral Engines (fictional automobile assembler)
A/P	Accounts Payable
Andon	(Japanese) signal light, as in andon display board
Big Three	The three largest automobile assemblers in the United States (GM, Ford and Chrysler)
BIQ	Built-in-Quality

CEO	Chief Executive Officer; the highest level executive in an organization outside of the Board of Directors
CIM	Computer Integrated Manufacturing
DM	Decimal minutes (certain stopwatches use this 1/10 of a minute scale)
DOWNTIME	Lost time due to machine breakdowns; also an acronym spelling out the eight categories of waste
E-stop	A control button or device that allows a worker to safely stop a machine, conveyor or other automated device
FF	Future Factory team
Five Why's	A problem-solving analysis method to get at an underlying root cause
FMEA	Failure Modes and Effects Analysis
Gemba	(Japanese) The place where work occurs
GUMBI	Great United Motor Builders, Inc. (fictional automobile assembler in Demot, CA—also fictional)
Hansei	(Japanese) Purposeful reflection with the purpose of improvement
House Model	Toyota house model for implementation made up of a foundation, pillars and roof and containing TPS principles
HR	Human Resources (department)
IE	Industrial Engineering
JCI	Johnson Controls, Inc. (specifically the automotive division)
Jidoka	Putting smart controls in the machine to stop the process when defects occur, so that the worker can avoid watching an automatically functioning machine
JIT	Just-in-Time (usually corresponding to production)
Jonah	A Theory of Constraints expert
Kanban	(Japanese) Signal card; used to pull materials to a process only when needed, not before
Keiretsu	(Japanese) A customer family with loyal suppliers for life
LOLE	Law of Least Effort (workers will find a way to do work that takes them less effort)
MIT	Massachusetts Institute of Technology
MRP	Material Requirements Planning
Murphy's Law	A pessimistic expectation meaning that anything that could go wrong usually will go wrong
OEE	Overall Equipment Effectiveness
OEM	Original Equipment Manufacturers (automobile assemblers)
OPT	Optimum Production Timetable (production scheduling software)
OR	Operations Research
PDCA	Plan, Do, Check and Act; systematic 4-step problem solving process and cycle of improvement
PM	Plant Manager throughout this book; also used once for Preventive Maintenance
Poka-Yoke	(Japanese) Devices or improvements used to error proof a process
SMED	Single Minute Exchange of Dies (quick changeover or setups)

SPC   Statistical Process Control
TPM   Total Productive Maintenance
TPS   Toyota Production System
TQM   Total Quality Management
Takasui   (Fictional) Consulting company
UM   University of Michigan
VP   Vice President
VSM   Value Stream Mapping
WIP   Work in Process (inventory)
WRI   Working Relations Index (relationship index of suppliers to their OEMs
           researched by Dr. John Henke)

Recommended for Further Reading

Imai, Masaki. *Gemba Kaizen: A Commonsense, Low-Cost Approach to Management.* New York: McGraw-Hill, 1997.

Liker, Jeffrey. (Ed.). *Becoming Lean: Inside Stories of U.S. Manufacturers.* Portland, OR: Productivity Press, 1997.

Liker, Jeffrey. *The Toyota Way: 14 Management Principles from the World's Greatest Manufacturer.* New York: McGraw-Hill, 2004.

Monden, Yasuhiro. *Toyota Production System: An Integrated Approach to Just-in-Time,* Third Edition. Norcross, GA: Engineering Management Press, 1998.

Ohno, Taiichi. *The Toyota Production System: Beyond Large-Scale Production.* New York: Productivity Press, 1988.

Rother, Michael and Rick Harris. *Creating Continuous Flow: An Action Guide for Managers, Engineers and Production Associates.* Brookline, MA: Lean Enterprise Institute, 2001.

Rother, Michael and John Shook. *Learning to See: Value-Stream Mapping to Add Value and Eliminate Muda.* Brookline, MA: Lean Enterprise Institute, 1999.

Shingo, Shigeo. *A Study of the Toyota Production System from an Industrial Engineering Viewpoint,* English translation (Andrew Dillon), Portland, OR: Productivity Press, 1989.

Spear, Steven and H. Kent Bowen. "Decoding the DNA of the Toyota Production System," *Harvard Business Review,* Sept.–Oct. 1999.

Suzaki, Kiyoshi. *The New Manufacturing Challenge: Techniques for Continuous Improvement.* New York: The Free Press, 1987.

Womack, James P. and Daniel T. Jones. *Lean Thinking: Banish Waste and Create Wealth in Your Corporation,* Second Edition. New York: Simon & Schuster, 2003.

Womack, James P., Daniel T. Jones, and Daniel Roos. *The Machine That Changed the World: The Story of Lean Production.* New York: HarperPerennial, 1991.

Index

About the Author

Steve Hoeft, a Toyota-trained Toyota Production System Coach at Altarum Institute, has helped organizations win over a dozen Shingo prizes. He is a practitioner, teacher, change agent and thought leader in applying lean principles widely and deeply to unique, knowledge-worker processes for hundreds of clients in multiple industries and application areas, including healthcare, new product development, supply chains, defense, government, and manufacturing.

While at a Big 3 automotive firm, Mr. Hoeft was trained by Eli Goldratt in the theory of constraints-based scheduling package OPT. His deep lean experience began at Delta (Kogyo) USA under a true sensei. Mr. Hoeft then continued to learn about lean systems when he moved to Johnson Controls, Inc., and studied directly with Toyota in Georgetown, KY. At JCI, Steve coauthored the *JCI Manufacturing System* (an ASTD Training Package of the Year award winner) and was responsible for starting the implementation process across all JCI plants. He was also a lean coach and consultant with Optiprise, Inc.

Mr. Hoeft holds a dual bachelor of science degree in industrial engineering and operations research from Wayne State University and an MBA from the University of Toledo. He is a Certified Project Manager Professional (PMP) through the Project Management Institute and a key instructor for the University of Michigan College of Engineering's *Lean Healthcare, Lean Manufacturing,* and *Lean Product Design* Certificate programs.